D1714044

Classical Equilibrium Statistical Mechanics

Classical Equilibrium Statistical Mechanics

COLIN J. THOMPSON

Mathematics Department,
University of Melbourne,
Parkville,
Victoria,
Australia

CLARENDON PRESS · OXFORD
1988

Oxford University Press, Walton Street, Oxford OX2 6DP
Oxford New York Toronto
Delhi Bombay Calcutta Madras Karachi
Petaling Jaya Singapore Hong Kong Tokyo
Nairobi Dar es Salaam Cape Town
Melbourne Auckland
and associated companies in
Berlin Ibadan

Oxford is a trade mark of Oxford University Press

Published in the United States
by Oxford University Press, New York

British Library Cataloguing in Publication Data
Thompson, Colin J.
Classical equilibrium statistical mechanics
1. Statistical mechanics
I. Title
530.1'32 QC 174.86
ISBN 0–19–851984–2

Library of Congress Cataloging in Publication Data
Thompson, Colin J.
Classical equilibrium statistical mechanics/Colin J. Thompson.
p. cm. Bibliography: p. Includes index.
1. Statistical mechanics. 2. Matter—Properties. I. Title.
QC174.8.T49 1988
530.1'32—dc19 87–28730 CIP
ISBN 0–19–851984–2

Typeset by Macmillan India Ltd, Bangalore 25

Printed in Great Britain
at the University Printing House, Oxford
by David Stanford
Printer to the University

PREFACE

This book is intended as an introductory text in statistical mechanics. The only prerequisites are advanced calculus and an elementary knowledge of probability theory, linear algebra, and classical mechanics. No prior knowledge of thermodynamics or statistical mechanics is assumed.

Statistical mechanics covers a wide variety of topics and impinges on many fields of science. Treatments of the subject range from the very physical to the very mathematical. The presentation here, while slightly biased in the mathematical or rigorous direction, does not lose sight of the physics. In order to keep the book to a manageable and readable size the topics covered have been confined to equilibrium properties of classical systems. The foundations of the subject are, however, presented in some detail and in certain areas the reader is brought to the threshold of current research activity, especially through the many problems and worked solutions that follow each chapter.

The aim of classical equilibrium statistical mechanics is to relate the microscopic properties of matter to the macroscopic, or bulk, behaviour of systems by providing a statistical procedure for calculating equilibrium thermodynamic quantities of systems whose microscopic behaviour is governed by the laws of classical mechanics. The statistical procedure, due to Gibbs, is presented in Chapter 2 where it is shown that the derived quantities satisfy the prerequisite laws of thermodynamics. These laws are discussed briefly in Chapter 1 along with thermodynamic relations of quantities which are derived from the Gibbs procedure.

Model systems are introduced in Chapter 3 and several one-dimensional examples of fluid and magnetic systems are treated exactly according to the Gibbs prescription. The thermodynamic limit, which is required for equivalent descriptions by the various Gibbs distributions, as well as for a rigorous basis for the study of phase transitions, is also presented in Chapter 3 for a discrete magnetic system as well as for a continuum fluid system.

The classical theories of phase transitions due to van der Waals, and Curie and Weiss are discussed in Chapter 4 along with their range of validity. The generally accepted characterization of phase transitions by critical exponents is also presented in this chapter along with the Yang–Lee theory of condensation and the Peierls argument for the existence of the phase transition for the two-dimensional Ising model.

Chapter 5 discusses the classical Ornstein–Zernike theory of correlations, fluctuation relations for fluid and magnetic systems, the decay of correlations, and the Griffiths spin-correlation inequalities.

Some exactly solved models which exhibit phase transitions are given in Chapter 6. These include Onsager's celebrated solution of the two-dimensional Ising model in zero field and the more recent eight-vertex and hard-hexagon models in two dimensions due to Baxter. The Gaussian and spherical models are also included as examples that can be solved exactly in arbitrary spatial dimension. The emphasis in this chapter is on methods and results rather than on technical mathematical details.

The final chapter discusses scaling theory and the renormalization group from a precise mathematical standpoint. Scaling theory provides relations between critical exponents whereas the renormalization-group method provides a means for calculating exponents from renormalization-group transformations. The renormalization-group method, or recipe, can be made fairly precise but in executing the recipe one usually has to make some approximations. Some examples of renormalization-group transformations are presented in the final section of the book.

It is a pleasure finally, to thank the numerous friends, colleagues, and students who have contributed their comments, thoughts, and ideas over the years. In particular, I have had the good fortune to have as a close associate and long-time friend, Mark Kac, to whose memory this book is dedicated.

Tony Guttmann, Doochul Kim, and Paul Pearce kindly read the entire manuscript and contributed many useful comments, suggestions, and corrections. To them I am particularly grateful.

Special thanks are due to Marjorie Funston for her skilful typing and editing of the manuscript, for the original drawings, and especially for her patience and understanding during the writing and preparation of the manuscript. This book is hers.

Melbourne C.J.T.
February 1987

CONTENTS

This book is for
Marjorie
and is dedicated to
the memory of
Mark Kac

1

THERMODYNAMICS

1.1 Introduction

The main aim of statistical mechanics is to derive macroscopic properties of matter from the laws governing the microscopic behaviour of the individual particles.

Thermodynamics consists of laws governing the relationships between macroscopic variables and as such is a necessary prerequisite for statistical mechanics. From a mathematical point of view the subject matter of thermodynamics has the advantage that it can be completely axiomatized. Since our main concern here, however, is statistical mechanics, we will present only the 'primitive concepts' and 'laws' and, for later reference, some particular thermodynamic relations for fluid and magnetic systems. A detailed account of the axiomatization of thermodynamics, which was due mainly to Carathéodory (1909) can be found in Buchdahl (1966). More extensive treatments of the subject than ours have been given by Callen (1960) and Pippard (1957).

The primitive concepts of thermodynamics are obvious undefined things, analogous to 'point' and 'line' in Euclidean geometry, such as:

1. '*System*': for example, a gas or a magnet.
2. '*State of a system*': specified, for example, by pressure P and volume V for a gas and by magnetic field H and magnetization M for a magnet.
3. '*Thermal equilibrium of a system*': a state of a system that does not change with time.
4. '*Equation of state*': an equation relating P, V, and temperature T for a gas or H, M, and T for a magnet. For example, the equation of state for an ideal gas of N particles is $PV=NkT$ where k is Boltzmann's constant and so forth. It should be clear in the following sections which concepts are primitive and which are not.

We now discuss the various laws of thermodynamics.

1.2 Empirical temperature and the zeroth law

As the name suggests, the zeroth law was added as an afterthought. Its main point is to make the concept of thermal equilibrium more precise and to introduce the notion of temperature.

For simplicity, we will consider here only one-component systems whose states are specified by a value of the pressure P and a value of the volume V. That is, a state of the system, which we take to be a gas, is specified by a point in the first quadrant of the (P, V)-plane. We could equally well consider a magnetic system with states specified by magnetic field H and magnetization M, the analogous thermodynamic variables to P and V, respectively. We will have more to say about this analogy later. More complicated systems, whose states are specified by more than two variables, can be treated in a similar fashion. (See, for example, Callen 1960.)

We now introduce the concept of equilibrium of two systems through the following

Postulate *Associated with a pair of system* A *and* B *there is a function* $f_{AB}(P_1, V_1; P_2, V_2)$ *of the states* (P_1, V_1) *of* A *and* (P_2, V_2) *of* B *such that* A *and* B *are in equilibrium if and only if*

$$f_{AB}(P_1, V_1; P_2, V_2)=0. \tag{1.2.1}$$

Note that this postulate implies that the states of two systems in equilibrium cannot be specified arbitrarily.

We can now state the

Zeroth law of thermodynamics *A state of equilibrium exists and is transitive. That is, if a system* A *is in equilibrium with a system* B *and* B *is in equilibrium with a system* C, *then* A *is in equilibrium with* C.

In view of eqn (1.2.1) transitivity means that

$$f_{AB}(P_1, V_1; P_2, V_2)=0 \qquad \text{and} \qquad f_{BC}(P_2, V_2; P_3, V_3)=0$$

imply that

$$f_{AC}(P_1, V_1; P_3, V_3)=0. \tag{1.2.2}$$

Transitivity allows us to define a temperature scale, specifically *empirical temperature*, associated with a given system. An absolute scale of temperature will follow later as a consequence of the first and second laws.

To introduce the notion of temperature we consider two systems A and C in equilibrium with given 'test system' B which may be thought of as a thermometer. We assume that for such a test system the first two equations of eqn (1.2.2) can be 'solved' for P_2 in terms of P_1, V_1, V_2 and P_3, V_3, V_2, respectively. Granted the existence of such a test system B, there then exist functions Θ_A and Θ_C such that

$$P_2=\Theta_A(P_1, V_1; V_2)=\Theta_C(P_3, V_3; V_2) \tag{1.2.3}$$

which is equivalent, from the zeroth law, to saying that A and C are in equilibrium.

Now, since B is a fixed system we can ignore the V_2 (and B) dependence of Θ_A and Θ_C. The quantity $\Theta_A(P_1, V_1)$ is then defined to be the empirical temperature of the system A, *as measured by the test system or thermometer* B.

A necessary and sufficient condition then for two systems to be in equilibrium is that they have the same empirical temperature as measured by a fixed test system or thermometer.

It is important to note that equilibrium via a test system is necessary for a proper definition of empirical temperature. A common error in the literature states that, since eqn (1.2.3) written as

$$\Theta_A(P_1, V_1; V_2) - \Theta_C(P_3, V_3; V_2) = 0 \tag{1.2.4}$$

is equivalent to the third equation of eqn (1.2.2), the left-hand side of eqn (1.2.4) must be independent of V_2 and, hence, Θ_A and Θ_C must be of the form

$$\begin{aligned} &\Theta_A(P_1, V_1; V_2) = \theta_A(P_1, V_1) + \psi(V_2) \\ \text{and} \quad &\\ &\Theta_C(P_3, V_3; V_2) = \theta_C(P_3, V_3) + \psi(V_2). \end{aligned} \tag{1.2.5}$$

The quantity $\theta_A(P_1, V_1)$ is then defined to be the empirical temperature of A. The error, of course, is that Θ_A and Θ_C need not have the form given in eqn (1.2.5).

Finally, we note that, if $\theta = \Theta(P, V)$ is the empirical temperature of a system and if we can invert the function Θ to obtain $P = P(V, \theta)$, we have the equation of state of the system expressed in the standard form of pressure as a function of volume and temperature. One of the aims of statistical mechanics is to derive an analytical expression for the equation of state. This is clearly beyond the scope of thermodynamics.

With a formulation of the notion of temperature we now move on to the first law.

1.3 Energy conservation and the first law

Before stating the first law we need to define some terms.

1. *Isotherms.* The states (P, V) of a system A with constant empirical temperature $\Theta_A(P, V)$ form a family of curves in the (P, V)-plane called isotherms. Thus through each point in the (P, V)-plane there is one and only one isotherm.

2. *Quasi-static process.* A change of state that takes place so slowly that the system can be considered to be in thermal equilibrium (with a test system) at each step of the change is called a quasi-static process.

3. *Reversible process.* A process is called reversible if the initial state can be obtained from the final state by following the inverse of each step of the original process.

4. *Adiabatic process.* A change of state that occurs while the system is thermally isolated (i.e. not in contact or in equilibrium with any other system) is called an adiabatic process.

5. *Work.* If a gas changes quasi-statically from one state to another along a curve Γ in the (P, V) plane, the work done by the system in the process is defined to be

$$\Delta W = \int_\Gamma P \mathrm{d}V. \tag{1.3.1}$$

If the process is also reversible it follows that the work done by the system in traversing the inverse path Γ^{-1} is the negative of eqn (1.3.1).

We can now state the

First law of thermodynamics

1. *Associated with each system* A *there is a function* U_A *of the state of the system, called the internal energy of* A, *such that the work done by the system in going from one state* (P_1, V_1) *to another state* (P_2, V_2) *in any adiabatic process is*

$$\Delta W = U_A(P_1, V_1) - U_A(P_2, V_2) \equiv -\Delta U_A \tag{1.3.2}$$

independently of the path joining the initial and final states. It follows from eqn (1.3.1) *that if the process is also quasi-static,* $\int_\Gamma P\mathrm{d}V$ *is independent of the path* Γ.

2. *The quantity of heat* Q_A *absorbed by the system* A *in any change of state is defined by*

$$Q_A = \Delta U_A + \Delta W \tag{1.3.3}$$

and, for an infinitesimal quasi-static change,

$$\delta Q = \mathrm{d}U + P\mathrm{d}V \tag{1.3.4}$$

where we have written δQ, *since the right-hand side of eqn* (1.3.4) *is not necessarily an exact differential of some function.*

Note that for a quasi-static process,

$$\begin{aligned}\Delta U &= U(P_2, V_2) - U(P_1, V_1)\\ &= \int_\Gamma \mathrm{d}U\\ &= \int_\Gamma (\delta Q - P\mathrm{d}V)\end{aligned} \tag{1.3.5}$$

where Γ is any path joining the initial state (P_1, V_1) and the final state (P_2, V_2). It follows that in any quasi-static cyclic process, where the final state is the same as the initial state, there is no change in the internal energy of the system even though heat may be absorbed and work may be done by the system. In a sense then, the first law is just a statement of *energy conservation.*

It is interesting to trace the history of heat and energy conservation to Clausius's statement (Clausius 1850) of the first and second laws of thermodynamics. Heat, as everyone knows, is a form of energy, but even as late as the middle of the last century it was almost universally considered as a substance, commonly known as caloric, and definitely not as a form of energy. Rumford (1798) was probably the first to attempt an experiment, in his typically flamboyant fashion utilizing the services of horses and cannons, to show that mechanical energy can be transformed into heat. His experiment simply showed that when a cannon was bored in water, the temperature of the water increased. Davy (1799) performed similar, though less flamboyant, experiments with ice but, since neither Rumford nor Davy had a satisfactory alternative theory, the caloric theory survived. More decisive experiments by Mayer in 1842, and especially by Joule (1845), paved the way for Clausius's 'mechanical theory of heat' (Clausius 1850), in which the first and second laws were first clearly enunciated.

The second law, which is discussed in the next section, was actually formulated before Clausius by Carnot (1824). His arguments were based in part on the caloric theory, however, so his work was neglected somewhat by later workers.

The reader interested in more historical detail is referred to Roller (1950) and Brown (1950).

1.4 Entropy and the second law

The second law can be discussed in a number of different but essentially equivalent ways. We will proceed in a more or less standard and historical way via Carnot cycles, Kelvin's statement of the second law (which is equivalent to Clausius's earlier statement given in Problem 1), the absolute-temperature scale, and entropy.

We begin by defining some terms.

1. *Adiabatic curves.* Consider a quasi-static adiabatic process for which, from eqn (1.3.4)

$$\delta Q = \mathrm{d}U + P\,\mathrm{d}V = 0,$$

or, equivalently, using the fact that $U = U(P, V)$,

$$\left(P + \frac{\partial U}{\partial V}\right)\mathrm{d}V + \frac{\partial U}{\partial P}\mathrm{d}P = 0. \tag{1.4.1}$$

The solution curves of the first-order differential equation (1.4.1) are called *adiabatic curves.*

We will *assume* that $U(P, V)$ is sufficiently regular so that there is one and only one adiabatic curve passing through a given point (P_0, V_0). In other words, given some state (P_0, V_0) we can find another state (P_1, V_1) which can be connected to (P_0, V_0) by an adiabatic curve.

2. *Heat bath.* An object that we will frequently use is a heat bath, which is defined to be a system so large that its empirical temperature remains constant when a finite amount of heat is added to it or subtracted from it.

3. *Carnot cycle.* For a particular system, a *Carnot cycle* is defined to be the *reversible* cyclic process shown schematically in Fig. 1.1.

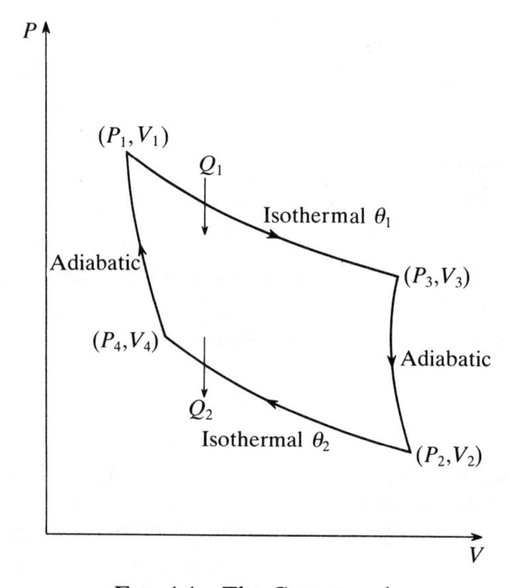

FIG. 1.1 The Carnot cycle.

Specifically, we start from a state (P_1, V_1) with empirical temperature $\theta_1 = \Theta(P_1, V_1)$ and expand the system isothermally to the state (P_3, V_3), keeping it in thermal equilibrium with a heat bath at constant temperature θ_1. During this part of the cycle a certain amount of heat Q_1 is absorbed by the system (from the heat bath). Next we expand the system adiabatically to (P_2, V_2) where the system has temperature $\theta_2 = \Theta(P_2, V_2)$. We then compress the system isothermally while it is in thermal equilibrium with a heat bath at temperature θ_2 until it reaches a state (P_3, V_3), from which the system can then be compressed adiabatically back to the initial state (P_1, V_1) of the cycle. During the isothermal compression, an amount of heat Q_2 is absorbed by the heat bath.

Assuming that all processes in the cycle are performed quasi-statically, the first law implies that

$$Q_1 = \int_{V_1}^{V_3} P(V, \theta_1)\mathrm{d}V + E(V_3, \theta_1) - E(V_1, \theta_1) \tag{1.4.2a}$$

and

$$-Q_2 = \int_{V_2}^{V_4} P(V, \theta_2)\mathrm{d}V + E(V_4, \theta_2) - E(V_2, \theta_2) \tag{1.4.2b}$$

along the two isotherms of the cycle, and that

$$0 = \int_{V_3}^{V_2} P\mathrm{d}V + E(V_2, \theta_2) - E(V_3, \theta_1) \tag{1.4.2c}$$

and

$$0 = \int_{V_4}^{V_1} P\mathrm{d}V + E(V_1, \theta_1) - E(V_4, \theta_2) \tag{1.4.2d}$$

along the two adiabatic portions of the cycle. In eqns (1.4.2) we have used the relation $\theta = \Theta(P, V)$ to express the pressure as $P = P(V, \theta)$ and the internal energy as

$$U(P, V) = U(P(V, \theta), V) \equiv E(V, \theta). \tag{1.4.3}$$

If we denote the Carnot cycle by C and add eqns (1.4.2) we deduce that the total work W done by the gas during the cycle is

$$W = \oint_C P\mathrm{d}V = Q_1 - Q_2. \tag{1.4.4}$$

We now define the efficiency e of the process described above to be the ratio of the work done by the gas to the quantity of heat taken in during the process, i.e.

$$e = \frac{W}{Q_1} = 1 - \frac{Q_2}{Q_1}. \tag{1.4.5}$$

We will show in a moment that the second law of thermodynamics implies that e is independent of the particular system undergoing the process and that as a result an absolute scale of temperature can be defined.

We start with

Kelvin's statement of the second law *No cyclic process exists whose sole effect is to extract heat from a substance and convert it entirely to work. In other words, there can be no perpetual-motion machine which converts the whole internal energy of a body into mechanical work.*

An equivalent, though less convenient statement for our present purposes, was given earlier by Clausius and is presented as Problem 1 at the end of this chapter.

We will now deduce from Kelvin's statement that the efficiency e of a Carnot cycle depends only on θ_1 and θ_2 and not on the particular system undergoing the cycle.

Thus consider two systems A and A′ undergoing Carnot cycles between temperatures θ_1 and θ_2. We assume that the relative size of A and A′ is such that $Q_1 = Q_1'$ (this is not necessary but it simplifies the argument considerably). We now show that $Q_2 = Q_2'$ or, equivalently, $W = W'$. Suppose that this is not so and that $W < W'$ or, equivalently, $Q_2 > Q_2'$. Consider now the combined process obtained by running the Carnot cycle for A′ followed by the *reversed* Carnot cycle for A. This process absorbs a positive amount of heat $Q_2 - Q_2'$ at temperature θ_2 and converts it entirely into work. This contradicts Kelvin's statement and hence $Q_2 \ngtr Q_2'$. Reversing the roles of A and A′ shows that $Q_2' \ngtr Q_2$ and, hence, $Q_2 = Q_2'$.

It follows immediately from the above observation that the ratio

$$\frac{Q_2}{Q_1} = f(\theta_1, \theta_2) \tag{1.4.6}$$

is a universal function of θ_1 and θ_2. We now show that $f(\theta_1, \theta_2)$ must be of the form

$$f(\theta_1, \theta_2) = \frac{\Phi(\theta_2)}{\Phi(\theta_1)}. \tag{1.4.7}$$

To prove this we again consider two Carnot cycles. The first operates between temperatures θ_1 and θ_2, absorbing a quantity of heat Q_1 during the first isothermal change and yielding a quantity of heat Q_2 during the second isothermal change. The second cycle operates between temperatures θ_2 and θ_3, absorbing a quantity of heat Q_2 and yielding a quantity of heat Q_3. For these two cycles, eqn (1.4.6) gives

$$\frac{Q_2}{Q_1} = f(\theta_1, \theta_2) \tag{1.4.8}$$

and

$$\frac{Q_3}{Q_2} = f(\theta_2, \theta_3). \tag{1.4.9}$$

For the combined cycle,

$$\frac{Q_3}{Q_1} = f(\theta_1, \theta_3). \tag{1.4.10}$$

Multiplying eqns (1.4.8) and (1.4.9) gives, from eqn (1.4.10)

$$f(\theta_1, \theta_3) = f(\theta_1, \theta_2) f(\theta_2, \theta_3) \tag{1.4.11}$$

and eqn (1.4.7) follows immediately. The function Φ in eqn (1.4.7) is then some universal function of the empirical temperature θ. The *absolute temperature* T defined by

$$T = \alpha\Phi(\theta) \tag{1.4.12}$$

where α is some constant scale factor, is then *universal*. It follows from eqns (1.4.6), (1.4.7), and (1.4.12) that the efficiency of a Carnot cycle for any system operating between absolute temperatures T_1 and T_2 is given by

$$e = 1 - \frac{Q_2}{Q_1} = 1 - \frac{T_2}{T_1}. \tag{1.4.13}$$

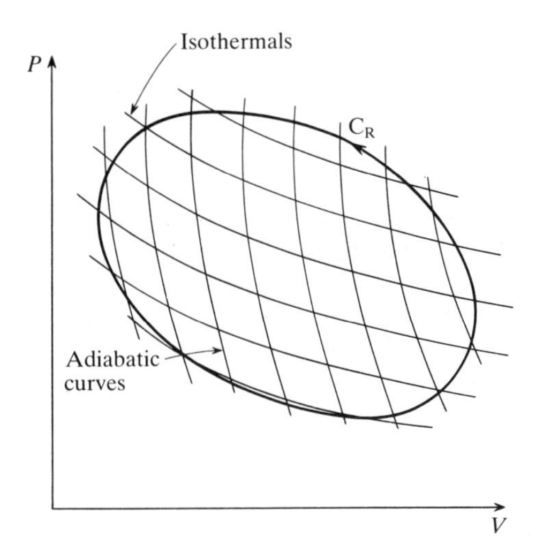

FIG. 1.2. Approximation of a closed reversible cycle C_R by a network of Carnot cycles.

Let us consider now an arbitrary closed reversible cycle C_R in the (P, V)-plane, and approximate it by a network of Carnot cycles, as shown in Fig. 1.2. For any Carnot cycle we have from eqn (1.4.13) that

$$\frac{Q_1}{T_1} + \frac{-Q_2}{T_2} = 0 \tag{1.4.14}$$

which says that the sum of (possibly negative) amounts of heat absorbed by the system, divided by the absolute temperature at which they are absorbed is zero. Applying this result to each little Carnot cycle in the network approximating C_R and allowing the network to become finer and finer, we obtain in the limit

Carnot's theorem *For any closed reversible cycle* C_R,

$$\oint_{C_R} \frac{\delta Q}{T} = 0. \tag{1.4.15}$$

This theorem implies that

$$\frac{\delta Q}{T} = dS \tag{1.4.16}$$

is an exact differential of a function

$$S = S(V, T) \tag{1.4.17}$$

of the state of the system, called the *entropy*. It is clear that S defined in this way is only determined up to an additive constant. In other words, only the *difference* in entropy between two states defined by

$$S(V_2, T_2) - S(V_1, T_1) = \int_\Gamma \frac{\delta Q}{T} \tag{1.4.18}$$

is defined, for Γ any *reversible* path joining the two states.

Let us consider now what happens to the above results if the paths in question are *not necessarily reversible*. In particular, consider any cycle of the form shown in Fig. 1.1 operating between two heat baths at absolute temperatures T_1 and T_2 absorbing quantities of heat Q_1 and $-Q_2$, respectively, from the two heat baths. In addition, consider two Carnot cycles C_0 and C'_0 operating between temperatures T_0 and T_1, and T_0 and T_2, respectively, absorbing quantities of heat Q_0 and Q'_0 at T_0 and yielding quantities of heat Q_1 and $-Q_2$ at T_1 and T_2, respectively. From eqn (1.4.13) we have

$$\frac{Q_0}{Q_1} = \frac{T_0}{T_1} \quad \text{for cycle } C_0 \tag{1.4.19}$$

and

$$\frac{Q'_0}{-Q_2} = \frac{T_0}{T_2} \quad \text{for cycle } C'_0. \tag{1.4.20}$$

Consider now the combined cycle consisting of C followed by C_0 followed by C'_0. A net amount of heat

$$Q = Q_0 + Q'_0 \tag{1.4.21}$$

is absorbed by this system and is converted entirely into work. In order to avoid a contradiction with the second law, we require

$$Q \leqslant 0, \tag{1.4.22}$$

or, in other words, from eqns (1.4.19) and (1.4.20) assuming $T_0 > 0$,

$$\frac{Q_1}{T_1} + \frac{-Q_2}{T_2} \leqslant 0. \tag{1.4.23}$$

Equation (1.4.23) is to be compared with eqn (1.4.14), which holds for reversible processes. In fact, if the cycle C is reversible, we can reverse the combined cycle $C + C_0 + C_0'$ to get eqn (1.4.23) with the inequality reversed. Equation (1.4.14) and Carnot's theorem then follow. In the more general case, however, we can repeat the argument leading from eqn (1.4.14) to eqn (1.4.15) to obtain

Clausius's theorem *For any cycle* C

$$\oint_C \frac{\delta Q}{T} \leqslant 0 \tag{1.4.24}$$

with equality holding if the cycle is reversible.

As a corollary to Clausius's theorem we have the more general form of (1.4.18)

$$S(V_2, T_2) - S(V_1, T_1) \geqslant \int_\Gamma \frac{\delta Q}{T} \tag{1.4.25}$$

which holds for any path Γ from state (V_1, T_1) to state (V_2, T_2) with equality holding if Γ is reversible.

To prove this result, consider the cycle C composed of Γ and the inverse of any reversible path Γ_R joining state (V_1, T_1) to state (V_2, T_2) as shown in Fig. 1.3. Equation (1.4.18) gives

$$S(V_2, T_2) - S(V_1, T_1) = \int_{\Gamma_R} \frac{\delta Q}{T}. \tag{1.4.26}$$

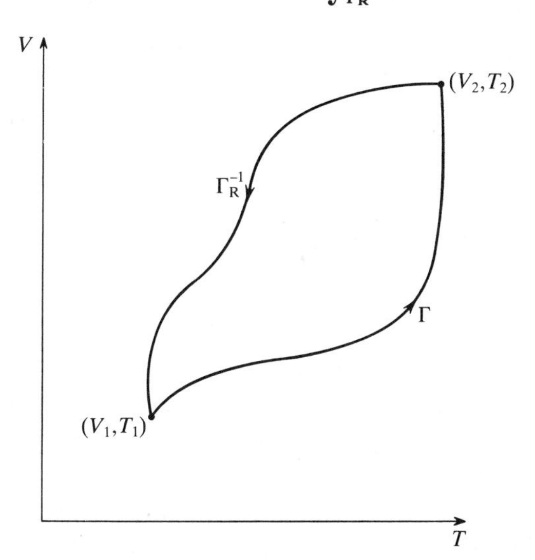

FIG. 1.3 Reversible path Γ_R and non-reversible path Γ connecting states '1' and '2'.

On the other hand, from Clausius's theorem eqn (1.4.24), we have

$$\oint_C \frac{\delta Q}{T} = \int_\Gamma \frac{\delta Q}{T} - \int_{\Gamma_R} \frac{\delta Q}{T} \leqslant 0. \tag{1.4.27}$$

The required result, eqn (1.4.25), then follows from eqns (1.4.26) and (1.4.27).

We note in particular from eqn (1.4.25) that, for a thermally isolated system where $\delta Q = 0$ for any process, entropy never decreases.

We summarize the foregoing as an

Alternative statement of the second law

1. *For* reversible *changes of state, there exists an absolute scale of temperature T such that $\delta Q/T$ is an exact differential of a function S, called entropy, of the state of the system.*

2. *For irreversible changes of state in a thermally isolated system, the entropy never decreases.*

Considerable care should be taken in interpreting part 2 of this version of the second law. The simple statement 'entropy never decreases' has caused much confusion and lively debate over the years, and needs to be very carefully qualified. The confusion usually arises when time enters into the discussion as a variable. It is natural to assign an interval of time to processes involving 'changes of state' even though, strictly speaking, we are dealing here with situations in which our systems are always in a state of 'time-independent' equilibrium. That is, we are considering idealized quasi-static processes which take place 'so slowly' that the system can always be considered to be in equilibrium.

The theory presented here then, is an equilibrium theory so time t does not enter as a variable. It is therefore *incorrect* to interpret part 2 of the above statement as implying

$$\frac{\partial S}{\partial t} \geqslant 0. \tag{1.4.28}$$

The correct way to interpret the statement is to say that if a state (V_2, T_2) can be reached (quasi-statically) from a state (V_1, T_1) by a reversible path and an irreversible path denoted, respectively, by Γ_R and Γ in eqns (1.4.26) and (1.4.27), then

$$S(V_2, T_2) \geqslant S(V_1, T_1) \tag{1.4.29}$$

where $S(V, T)$ is the equilibrium entropy defined by eqn (1.4.18). A reasonable interpretation then would be to say that an equilibrium state of a system corresponds to a state of maximum entropy.

The definition of a non-equilibrium or time-dependent entropy and the validity of (1.4.28) is an entirely different question and one which is indeed hotly debated.

There is one more 'law' which came sometime after the first and second laws. It is sometimes called the *third law of thermodynamics* and sometimes *Nernst's theorem*, after Nernst, who first stated it in 1905. The statement is simply

The third law *The entropy of a system at absolute temperature zero is a universal constant which may be taken to be zero.*

As there are model systems in classical equilibrium statistical mechanics (e.g. the n-vector model for $n>1$) which violate the third law, one should interpret Nernst's statement as an empirical law based on physical observations rather than as a theorem in the strict mathematical sense. Alternatively, one could argue that quantum effects are important at low temperatures so that classical model systems may become unphysical as the temperature approaches absolute zero. Since we will not concern ourselves here with low-temperature physics, the interested reader is referred to Huang (1963), or any standard textbook on thermodynamics and statistical mechanics, for the basis and a discussion of the 'third law'.

1.5 Free energies and derived quantities for fluid and magnetic systems

Two state functions which are useful in determining the equilibrium state of a system are the Helmholtz free energy Ψ and the Gibbs free energy Φ defined, respectively, by

$$\Psi = U - TS \tag{1.5.1}$$

and

$$\Phi = U - TS + PV \tag{1.5.2}$$

for a gas or fluid system, or

$$\Phi = U - TS - HM \tag{1.5.3}$$

for a magnetic system where U, T, S, P, V, H, and M are, respectively, internal energy, absolute temperature, entropy, pressure, volume, magnetic field, and magnetization.

In general, equations appropriate for magnetic systems are obtained from corresponding equations for fluid systems by making the substitutions H for P and $-M$ for V. For simplicity, we will work mainly in this section with the fluid form of the equations.

The essential properties of Ψ and Φ, which are useful in determining the equilibrium state of a system, are contained in the following

Theorem

1. *For a mechanically isolated system at constant temperature, Ψ never increases.*
2. *For a system at constant pressure and temperature, Φ never increases.*

These results are usually interpreted as saying that the equilibrium states for such systems correspond to states of minimum Ψ and Φ, respectively.

In order to prove the first statement of the theorem we note from eqn (1.4.25) that, for an infinitesimal isothermal transformation,

$$\frac{\Delta Q}{T} \leqslant \Delta S. \tag{1.5.4}$$

From the first law eqn (1.3.3) we can rewrite eqn (1.5.4) as

$$\Delta W \leqslant -\Delta U + T\Delta S. \tag{1.5.5}$$

The right-hand side of eqn (1.5.5) is $-\Delta\Psi$, and for a mechanically isolated system $\Delta W = 0$. It follows that

$$\Delta\Psi \leqslant 0 \tag{1.5.6}$$

with equality holding if the transformation is reversible.

For a system with constant pressure as well as constant temperature, we have

$$\Delta W = P\Delta V \tag{1.5.7}$$

in addition to eqn (1.5.5). These two equations imply that

$$\Delta\Phi = P\Delta V + \Delta U - T\Delta S \leqslant 0 \tag{1.5.8}$$

which completes the proof of the theorem.

In order to derive thermodynamic quantities and relations we consider Ψ or Φ as functions of any two of the three variables P, V, T which are related to one another through the equation of state. For example, if we consider, as is usual $\Psi = \Psi(V, T)$, we find from eqn (1.5.1) that

$$\begin{aligned} \mathrm{d}\Psi &= \mathrm{d}U - T\mathrm{d}S - S\mathrm{d}T \\ &= \delta Q - P\mathrm{d}V - T\mathrm{d}S - S\mathrm{d}T \\ &= -P\mathrm{d}V - S\mathrm{d}T \end{aligned} \tag{1.5.9}$$

where use has been made of the infinitesimal forms, eqns (1.3.4) and (1.4.16), of the first and second laws, respectively. It follows immediately that

$$P = -\left(\frac{\partial\Psi}{\partial V}\right)_T \tag{1.5.10}$$

and

$$S = -\left(\frac{\partial\Psi}{\partial T}\right)_V \tag{1.5.11}$$

where, to be specific, we have indicated by a subscript the variable to be held fixed during the partial differentiation. Assuming continuity of partial derivatives, eqns (1.5.10) and (1.5.11) imply, on differentiating with respect to T and

V, that

$$\left(\frac{\partial P}{\partial T}\right)_V = \left(\frac{\partial S}{\partial V}\right)_T \tag{1.5.12}$$

which is one of a class of relations known as *Maxwell's relations*. More are given as Problems at the end of this chapter.

The above formulae and relations are appropriate for a gas or fluid system. For a magnetic system we make the replacements H for P and $-M$ for V to obtain, in place of eqns (1.5.9) and (1.5.10)

$$\mathrm{d}\Psi = H\mathrm{d}M - S\mathrm{d}T \tag{1.5.13}$$

from which one obtains

$$H = \left(\frac{\partial \Psi}{\partial M}\right)_T \tag{1.5.14}$$

and

$$S = -\left(\frac{\partial \Psi}{\partial T}\right)_M. \tag{1.5.15}$$

Alternatively, if one starts from the Gibbs free energy eqn (1.5.3) and takes, as is usual, the field H and temperature T as the independent thermodynamic variables, one has

$$\mathrm{d}\Phi = -M\mathrm{d}H - S\mathrm{d}T \tag{1.5.16}$$

and hence

$$M = -\left(\frac{\partial \Phi}{\partial H}\right)_T \tag{1.5.17}$$

and

$$S = -\left(\frac{\partial \Phi}{\partial T}\right)_H. \tag{1.5.18}$$

The task of statistical mechanics is to actually derive explicit expressions for Ψ or Φ or any other appropriate free energy or potential. Quantities of physical interest are then obtained straightforwardly by differentiation. For example, the pressure is given by eqn (1.5.10) from which one can then obtain the *isothermal compressibility* K_T defined by

$$K_T^{-1} = -V\left(\frac{\partial P}{\partial V}\right)_T = V\left(\frac{\partial^2 \Psi}{\partial V^2}\right)_T. \tag{1.5.19}$$

Similarly, for a magnetic system, the magnetization is given by (1.5.17), from which one can then obtain the *isothermal susceptibility* χ_T defined by

$$\chi_T = \left(\frac{\partial M}{\partial H}\right)_T = -\left(\frac{\partial^2 \Phi}{\partial H^2}\right)_T. \tag{1.5.20}$$

Another quantity of physical interest is the specific heat at constant 'X', where X could denote, for example, P, V, H, or M, defined by

$$C_X = \left(\frac{\delta Q}{\mathrm{d}T}\right)_X = T\left(\frac{\partial S}{\partial T}\right)_X \tag{1.5.21}$$

where use has been made of the definition of entropy eqn (1.4.16).

We have for example, from eqns (1.5.11) and (1.5.18), that

$$C_V = -T\left(\frac{\partial^2 \Psi}{\partial T^2}\right)_V \tag{1.5.22}$$

and

$$C_H = -T\left(\frac{\partial^2 \Phi}{\partial T^2}\right)_H, \tag{1.5.23}$$

and so forth.

We note finally that certain physical restrictions on derived thermodynamic quantities impose mathematical restrictions on Ψ and Φ. For example, we require on physical grounds that K_T be non-negative. From eqn (1.5.19) this implies that

$$\left(\frac{\partial^2 \Psi}{\partial V^2}\right)_T \geqslant 0, \tag{1.5.24}$$

or, in other words, Ψ must be a *convex function* of volume. In addition, the requirement of non-negative pressure implies from eqn (1.5.10) that Ψ must be a *monotonic non-increasing function* of V. Thus as a function of V, Ψ must have the form shown in Fig. 1.4.

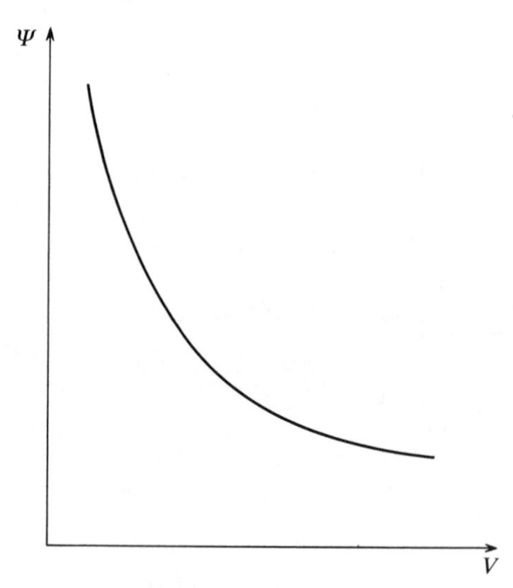

FIG. 1.4 The Helmholtz free energy $\Psi(V, T)$ is a convex monotonically decreasing function of V for fixed T.

Similarly, non-negative isothermal susceptibility implies, from eqn (1.5.20), that

$$\left(\frac{\partial^2 \Phi}{\partial H^2}\right)_T \leqslant 0, \tag{1.5.25}$$

which means that Φ must be a concave (or convex downward) function of H. Likewise, non-negative specific heats require, from eqns (1.5.22) and (1.5.23), that Ψ and Φ be concave functions of T.

These convexity properties of the free energies turn out to be important and will be discussed in more detail in later chapters.

1.6 Thermodynamic relations

Here we derive a particular thermodynamic relation connecting specific heats and certain derivatives of the magnetization. Other such relations are given as problems at the end of this chapter.

The particular relation we derive is

$$C_H - C_M = T\chi_T^{-1}\left(\frac{\partial M}{\partial T}\right)_H^2 \tag{1.6.1}$$

where the specific heats, defined by eqn (1.5.21), are given by

$$C_H = T\left(\frac{\partial S}{\partial T}\right)_H \quad \text{and} \quad C_M = T\left(\frac{\partial S}{\partial T}\right)_M, \tag{1.6.2}$$

and the isothermal susceptibility, from eqn (1.5.20), is given by

$$\chi_T = \left(\frac{\partial M}{\partial H}\right)_T. \tag{1.6.3}$$

In order to derive eqn (1.6.1) we note first of all, from eqns (1.5.17) and (1.5.18), the Maxwell relation

$$\left(\frac{\partial S}{\partial H}\right)_T = \left(\frac{\partial M}{\partial T}\right)_H. \tag{1.6.4}$$

Next, if we consider the entropy as a function of M and T, expressed in the form $S(H(M, T), T)$, we have from the chain rule that

$$\left(\frac{\partial S}{\partial T}\right)_M = \left(\frac{\partial S}{\partial T}\right)_H + \left(\frac{\partial S}{\partial H}\right)_T \left(\frac{\partial H}{\partial T}\right)_M. \tag{1.6.5}$$

Finally, since H, M, and T are connected by the equation of state, the chain relation (see Problem 4) gives

$$\left(\frac{\partial H}{\partial T}\right)_M = -\left(\frac{\partial M}{\partial T}\right)_H \Big/ \left(\frac{\partial M}{\partial H}\right)_T. \tag{1.6.6}$$

Substituting eqns (1.6.4) and (1.6.6) into eqn (1.6.5) and using the definitions (1.6.2) of C_H and C_M then gives the required relation, eqn (1.6.1).

The analogous relation for a fluid system is

$$C_P - C_V = TV\alpha^2 K_T^{-1} \tag{1.6.7}$$

where C_P and C_V are the specific heats at constant pressure and volume, respectively, K_T is the isothermal compressibility, and α, the *coefficient of thermal expansion*, is defined by

$$\alpha = \frac{1}{V}\left(\frac{\partial V}{\partial T}\right)_P. \tag{1.6.8}$$

The derivation of eqn (1.6.7) is similar to the derivation of eqn (1.6.1) and is left as an exercise for the reader. Other relations are given as problems at the end of this chapter.

At this stage the study of thermodynamics, at least from a mathematical point of view, becomes essentially an exercise in partial differentiation. This is then a good point at which to leave thermodynamics and move on to statistical mechanics.

1.7 Problems

1. Clausius's statement of the second law is the following: There is no transformation whose sole effect is to extract heat from a colder reservoir and transfer it to a hotter reservoir. Show that this statement is equivalent to Kelvin's statement.

2. By considering in turn the internal energy $U=U(V, S)$ as a function of volume and entropy, the Gibbs free energy $\Phi=\Phi(P, T)$ as a function of pressure and temperature, and the *enthalpy* $E=U+PV=E(P, S)$ as a function of pressure and entropy, deduce the following Maxwell relations:

$$\left(\frac{\partial T}{\partial V}\right)_S = -\left(\frac{\partial P}{\partial S}\right)_V; \quad \left(\frac{\partial S}{\partial P}\right)_T = -\left(\frac{\partial V}{\partial T}\right)_P; \quad \left(\frac{\partial V}{\partial S}\right)_P = \left(\frac{\partial T}{\partial P}\right)_S.$$

3. We defined a convex function $f(x)$ in the text by $\mathrm{d}^2f/\mathrm{d}x^2 \geqslant 0$, assuming that $f(x)$ is twice differentiable. This means that the function has the form shown in Fig. 1.4. Prove that the following obvious geometrical facts are implied by the statement $\mathrm{d}^2f/\mathrm{d}x^2 \geqslant 0$ for twice differentiable functions.

(a)
$$f\left(\frac{x_1+x_2}{2}\right) \leqslant \tfrac{1}{2}[f(x_1)+f(x_2)].$$

That is, the midpoint of the curve between $f(x_1)$ and $f(x_2)$ lies below the midpoint of the chord joining $f(x_1)$ and $f(x_2)$.

(b)
$$f'(x_1) \leqslant \frac{f(x_2)-f(x_1)}{x_2-x_1}.$$

That is, tangents to the curve lie below the curve.

4. Let x, y, and z be three variables satisfying a functional relationship $f(x, y, z)=0$. Deduce that

(a)
$$\left(\frac{\partial x}{\partial y}\right)_z=\left[\left(\frac{\partial y}{\partial x}\right)_z\right]^{-1};$$

(b)
$$\left(\frac{\partial x}{\partial y}\right)_z\left(\frac{\partial y}{\partial z}\right)_x\left(\frac{\partial z}{\partial x}\right)_y=-1;$$

and, if w is a function of any two of x, y, and z, that

(c)
$$\left(\frac{\partial x}{\partial y}\right)_w\left(\frac{\partial y}{\partial z}\right)_w=\left(\frac{\partial x}{\partial z}\right)_w.$$

5. If a gas has an equation of state of the form

$$PV=\phi(T) \qquad \text{(Boyle's law)}$$

with internal energy U a function of T only, i.e.

$$U=\psi(T) \qquad \text{(Guy-Lussac's and/or Joule's law)},$$

deduce from the first and second laws that the gas is an ideal gas, i.e. having equation of state

$$PV=RT$$

where R is a constant (called the *ideal gas constant*).

6. For an ideal gas show that

$$C_P-C_V=R.$$

7. By considering V and T as independent variables, deduce from the first and second laws that

$$T\mathrm{d}S=C_V\mathrm{d}T+T\left(\frac{\partial P}{\partial T}\right)_V\mathrm{d}V.$$

Obtain a similar relation involving C_P.

8. By considering adiabatic transformations ($\mathrm{d}S=0$) deduce from the results of Problem 7 that

$$C_V=\frac{TV\alpha^2K_S}{K_T(K_T-K_S)} \qquad \text{and} \qquad C_P=\frac{TV\alpha^2}{K_T-K_S}$$

where K_T and α are defined, respectively, by eqns (1.5.19) and (1.6.8) and K_S is the adiabatic compressibility defined by

$$K_S^{-1}=-V\left(\frac{\partial P}{\partial V}\right)_S.$$

1.8 Solutions to problems

1. To show that Clausius's statement (C) is equivalent to Kelvin's statement (K) we show that C false implies that K is false and vice versa. Suppose firstly that C is false.

That is, we can transfer a certain amount of heat Q_1 from a reservoir at temperature T_2 to a reservoir at a higher temperature T_1. Suppose now, with the aid of a Carnot cycle, we then absorb this amount of heat Q_1 and produce an amount of work W. Since the reservoir at temperature T_1 receives and yields the same amount of heat, it follows that the sole effect of this process is to extract heat from a reservoir and convert it entirely into work. This contradicts K and hence K is false.

Finally, let us suppose that K is false. That is, we can extract heat from a reservoir at temperature T_2 and convert it entirely into work. We can then convert this work into heat and transfer it to a reservoir at temperature $T_1 > T_2$. The sole effect of this process is to transfer an amount of heat from a colder reservoir to a hotter reservoir. This contradicts C and hence C is false.

The proof of equivalence of C and K is then complete.

2. (i)
$$\mathrm{d}U = \left(\frac{\partial U}{\partial S}\right)_V \mathrm{d}S + \left(\frac{\partial U}{\partial V}\right)_S \mathrm{d}V = T\mathrm{d}S - P\mathrm{d}V.$$

Hence,
$$T = \left(\frac{\partial U}{\partial S}\right)_V \quad \text{and} \quad P = -\left(\frac{\partial U}{\partial V}\right)_S.$$

Assuming continuity of partial derivatives, we then have
$$\left(\frac{\partial T}{\partial V}\right)_S = \frac{\partial^2 U}{\partial V \partial S} = -\left(\frac{\partial P}{\partial S}\right)_V.$$

(ii)
$$\mathrm{d}\Phi = \left(\frac{\partial \Phi}{\partial P}\right)_T \mathrm{d}P + \left(\frac{\partial \Phi}{\partial T}\right)_P \mathrm{d}T = V\mathrm{d}P - S\mathrm{d}T.$$

Hence,
$$V = \left(\frac{\partial \Phi}{\partial P}\right)_T \quad \text{and} \quad S = -\left(\frac{\partial \Phi}{\partial T}\right)_P.$$

Assuming continuity of partial derivatives, we then have
$$\left(\frac{\partial S}{\partial P}\right)_T = -\frac{\partial^2 \Phi}{\partial P \partial T} = -\left(\frac{\partial V}{\partial T}\right)_P.$$

(iii)
$$\mathrm{d}E = \left(\frac{\partial E}{\partial P}\right)_S \mathrm{d}P + \left(\frac{\partial E}{\partial S}\right)_P \mathrm{d}S = V\mathrm{d}P + T\mathrm{d}S.$$

Hence,
$$V = \left(\frac{\partial E}{\partial P}\right)_S \quad \text{and} \quad T = \left(\frac{\partial E}{\partial S}\right)_P.$$

Assuming continuity of partial derivatives, we then have
$$\left(\frac{\partial V}{\partial S}\right)_P = \frac{\partial^2 E}{\partial S \partial P} = \left(\frac{\partial T}{\partial P}\right)_S.$$

3. (a) Since $f''(x) \geqslant 0$, $f'(x)$ is monotonic increasing. Thus if $x_1 < x_2$,
$$f'(x) \leqslant f'\left(x + \frac{x_2 - x_1}{2}\right) \quad \text{for all } x.$$

Integrating this inequality with respect to x from x_1 to $(x_1+x_2)/2$, we obtain

$$f\left(\frac{x_1+x_2}{2}\right)-f(x_1)\leqslant f(x_2)-f\left(\frac{x_1+x_2}{2}\right)$$

which gives the required result.

(b) From the mean-value theorem there exists $x_1\leqslant\xi\leqslant x_2$ for $x_1<x_2$ such that

$$f'(\xi)=\frac{f(x_2)-f(x_1)}{x_2-x_1}.$$

By monotonicity of f', we have $f'(\xi)\geqslant f'(x_1)$ which gives the required result.

4. We assume that the first partial derivatives of f are non-zero so that the relationship $f(x, y, z)=0$ with z, in particular, fixed can be solved for $y=y(x)$.

(a) When z is fixed, we differentiate

$$f(x, y(x), z)=0$$

with respect to x to obtain

$$0=\left(\frac{\partial f}{\partial x}\right)_z=\frac{\partial f}{\partial x}+\frac{\partial f}{\partial y}\left(\frac{\partial y}{\partial x}\right)_z,$$

i.e.

$$\left(\frac{\partial y}{\partial x}\right)_z=-\frac{\partial f/\partial x}{\partial f/\partial y}. \tag{1}$$

Similarly, if we solve $f(x, y, z)=0$ with z fixed for $x=x(y)$, we obtain

$$0=\left(\frac{\partial f}{\partial y}\right)_z=\frac{\partial f}{\partial x}\left(\frac{\partial x}{\partial y}\right)_z+\frac{\partial f}{\partial y},$$

i.e.

$$\left(\frac{\partial x}{\partial y}\right)_z=-\frac{\partial f/\partial y}{\partial f/\partial x}. \tag{2}$$

The required result follows immediately from eqns (1) and (2).

(b) Using arguments similar to those in part (a), we obtain

$$\left(\frac{\partial y}{\partial z}\right)_x=-\frac{\partial f/\partial z}{\partial f/\partial y} \tag{3}$$

and

$$\left(\frac{\partial z}{\partial x}\right)_y=-\frac{\partial f/\partial x}{\partial f/\partial z}. \tag{4}$$

Multiplying eqns (2), (3), and (4) together gives the required result.

(c) If w is a function of any two of x, y, and z we can express the functional relationship $f(x, y, z)=0$ with w fixed in terms of any one of the three variables x, y, or z. Expressed as a function of y, for example, we obtain

$$0=\left(\frac{\partial f}{\partial y}\right)_w=\left(\frac{\partial f}{\partial x}\right)\left(\frac{\partial x}{\partial y}\right)_w+\frac{\partial f}{\partial y}+\left(\frac{\partial f}{\partial z}\right)\left(\frac{\partial z}{\partial y}\right)_w=0. \tag{5}$$

Similarly,

$$0=\left(\frac{\partial f}{\partial z}\right)_w=\left(\frac{\partial f}{\partial x}\right)\left(\frac{\partial x}{\partial z}\right)_w+\frac{\partial f}{\partial y}\left(\frac{\partial y}{\partial z}\right)_w+\frac{\partial f}{\partial z}=0. \tag{6}$$

Multiplying eqn (5) by $\left(\frac{\partial y}{\partial z}\right)_w$ and subtracting the resulting equation from eqn (6) gives

$$\frac{\partial f}{\partial x}\left[\left(\frac{\partial x}{\partial y}\right)_w\left(\frac{\partial y}{\partial z}\right)_w-\left(\frac{\partial x}{\partial z}\right)_w\right]=0$$

from which the required result follows (since we are assuming that $\partial f/\partial x\neq 0$).

5. If $PV=\phi(T)$ and $U=\psi(T)$, the first law applied to an infinitesimal change gives

$$\begin{aligned}\delta Q&=P\mathrm{d}V+\mathrm{d}U\\&=V^{-1}\phi(T)\mathrm{d}V+\frac{\mathrm{d}\psi}{\mathrm{d}T}\mathrm{d}T.\end{aligned}$$

Now from the second law $\delta Q/T$ is an exact differential; hence,

$$\frac{\partial}{\partial T}[(TV)^{-1}\phi(T)]=\frac{\partial}{\partial V}\left[T^{-1}\frac{\mathrm{d}\psi}{\mathrm{d}T}\right]=0,$$

since $T^{-1}(\mathrm{d}\psi/\mathrm{d}T)$ is a function only of T. It follows that

$$T^{-1}\phi(T)=R, \qquad \text{a constant}$$

and hence

$$PV=\phi(T)=RT.$$

6. From the first law,

$$\delta Q=P\mathrm{d}V+\mathrm{d}U,$$

we have, since U is a function only of T, that

$$C_P=\left(\frac{\delta Q}{\partial T}\right)_P=P\left(\frac{\partial V}{\partial T}\right)_P+\frac{\mathrm{d}U}{\mathrm{d}T}. \tag{1}$$

Similarly,

$$C_V=\left(\frac{\delta Q}{\partial T}\right)_V=\frac{\mathrm{d}U}{\mathrm{d}T}. \tag{2}$$

But, from the equation of state for an ideal gas,

$$P\left(\frac{\partial V}{\partial T}\right)_P=R. \tag{3}$$

The required result follows easily from eqns (1), (2), and (3).

7. If we consider V and T as independent variables we obtain from the first law

$$\delta Q = T\mathrm{d}S = \mathrm{d}U + P\mathrm{d}V$$
$$= \left(\frac{\partial U}{\partial T}\right)_V \mathrm{d}T + \left(\frac{\partial U}{\partial V}\right)_T \mathrm{d}V + P\mathrm{d}V$$
$$= C_V\mathrm{d}T + \left[\left(\frac{\partial U}{\partial V}\right)_T + P\right]\mathrm{d}V. \qquad (1)$$

Now, since $\mathrm{d}S$ is an exact differential,

$$\frac{\partial}{\partial V}(T^{-1}C_V) = \left\{\frac{\partial}{\partial T}\left[T^{-1}\left(\frac{\partial U}{\partial V}\right)_T + T^{-1}P\right]\right\}_V,$$

which gives

$$\left(\frac{\partial U}{\partial V}\right)_T + P = T\left(\frac{\partial P}{\partial T}\right)_V. \qquad (2)$$

Substituting eqn (2) into eqn (1) gives the required result.

Similarly, if P and T are taken as independent variables, we obtain

$$T\mathrm{d}S = C_P\mathrm{d}T - T\left(\frac{\partial V}{\partial T}\right)_P \mathrm{d}P.$$

8. For an adiabatic transformation,

$$0 = T\mathrm{d}S = C_P\mathrm{d}T - T\left(\frac{\partial V}{\partial T}\right)_P \mathrm{d}P$$

and

$$\mathrm{d}V = \left(\frac{\partial V}{\partial S}\right)_P \mathrm{d}S + \left(\frac{\partial V}{\partial P}\right)_S \mathrm{d}P = \left(\frac{\partial V}{\partial P}\right)_S \mathrm{d}P.$$

Since

$$\mathrm{d}T = \left(\frac{\partial T}{\partial V}\right)_P \mathrm{d}V + \left(\frac{\partial T}{\partial P}\right)_V \mathrm{d}P,$$

it follows that

$$C_P\left[\left(\frac{\partial T}{\partial V}\right)_P\left(\frac{\partial V}{\partial P}\right)_S + \left(\frac{\partial T}{\partial P}\right)_V\right] = T\left(\frac{\partial V}{\partial T}\right)_P.$$

The required result for C_P is then obtained from the results of Problem 1.4, that is

$$\left(\frac{\partial T}{\partial V}\right)_P = \left(\frac{\partial V}{\partial T}\right)_P^{-1}$$

and

$$\left(\frac{\partial T}{\partial P}\right)_V\left(\frac{\partial P}{\partial V}\right)_T\left(\frac{\partial V}{\partial T}\right)_P = -1.$$

The expression for C_V is obtained in a similar way.

2

THE GIBBS DISTRIBUTIONS

2.1 Introduction

The aim of statistical mechanics is to derive macroscopic properties of matter from the microscopic laws governing the behaviour of the individual particles. The macroscopic properties are expressed in terms of thermodynamic quantities discussed in Chapter 1. In statistical mechanics one must first define the thermodynamic variables and then check that the laws of thermodynamics expressed in terms of these variables are valid. Unlike thermodynamics, however, which consists only of laws and relations connecting thermodynamic quantities such as energy, entropy, specific heat, etc., statistical mechanics aims to derive explicit expressions for these quantities.

Before we can proceed, we must specify the microscopic description. Here we consider only classical systems of particles whose behaviour is governed by Newton's laws of motion. We will adopt the Hamiltonian formulation of these laws and represent a microscopic state of a system with n degrees of freedom by a point

$$(p, q)=(p_1, p_2, \ldots, p_n, q_1, q_2, \ldots, q_n) \tag{2.1.1}$$

in $2n$-dimensional *phase space*, or Γ-space, consisting of n canonical coordinates q_i and their conjugate momenta p_i, $i=1, 2, \ldots, n$. For example, if our system is three-dimensional, (q_1, q_2, q_3) and (p_1, p_2, p_3) would represent the position vector and momentum vector of a particle, (q_4, q_5, q_6) and (p_4, p_5, p_6) the position vector and momentum vector of another particle, and so forth, so if we have a three-dimensional system composed of N particles, $n=3N$ and our phase space is $6N$-dimensional.

The time evolution of a state of the system is governed by Hamilton's equations of motion

$$\frac{\mathrm{d}q_i}{\mathrm{d}t}=\frac{\partial \mathscr{H}}{\partial p_i}, \quad \frac{\mathrm{d}p_i}{\mathrm{d}t}=-\frac{\partial \mathscr{H}}{\partial q_i}, \qquad i=1, 2, \ldots, n \tag{2.1.2}$$

where $\mathscr{H}=\mathscr{H}(p, q)$ is the Hamiltonian of the system, i.e. the kinetic energy plus the potential energy of the system. We will consider throughout only Hamiltonians of the form

$$\mathscr{H}(p, q)=\sum_{i=1}^{n} \frac{p_i^2}{2m}+V(q_1, q_2, \ldots, q_n) \tag{2.1.3}$$

that do not depend explicitly on time. (Various restrictions on the potential-energy function V will be imposed later.)

For such systems the total energy is a conserved quantity or, in other words, a constant of the motion. Thus the motion of a system with total energy E is restricted to the $(2n-1)$-dimensional *energy surface*

$$\mathscr{H}(p, q)=E. \tag{2.1.4}$$

Hamilton's equations of motion (2.1.2) tell us how points move in phase space as time evolves. For Hamiltonians that do not depend on any time derivatives of p and q, such as eqn (2.1.3), Hamilton's eqns (2.1.2) constitute a set of first-order differential equations that are invariant under time reversal. It is clear also that, if the coordinates of a point in phase space are given at any time, eqns (2.1.2) determine the motion of the point *uniquely* for all other times.

2.2 Macroscopic variables and the approach to equilibrium

The difficulty in deducing equilibrium properties of a system from the microscopic behaviour of the individual particles lies in the fact that we are essentially dealing with two distinct levels of description. Equilibrium is observed only on the macroscopic level and is described by relatively few macroscopic variables. On the microscopic level, however, one describes the motion of all the particles. It is because of this difference in description that one hopes for a reconciliation between the time reversibility on the microscopic level and the apparent irreversibility on the macroscopic level.

To begin, we must say precisely what is meant by a macroscopic variable. Unfortunately, this notion is somewhat arbitrary and depends on how detailed a description one wants of the system. Gibbs suggests that a (small) finite number of macroscopic variables

$$y_i=f_i(p, q), \qquad i=1, 2, \ldots, m \tag{2.2.1}$$

of the microscopic state (p, q) of the system, used to describe the macroscopic state of a system, should satisfy the following conditions:

1. Each macroscopic state defined by a set of y_i values corresponds to a region of phase space, or more precisely, a portion of the energy surface $\mathscr{H}(p, q)=E$.
2. For a system composed of a large number N of particles there is one set of values $\bar{y}_1, \ldots, \bar{y}_m$ which corresponds to a region of the energy surface whose area is overwhelmingly the largest.

The first requirement corresponds to a 'coarse graining' of the energy surface and is essential for any macroscopic description. The amount of coarse graining clearly depends on the level of description demanded by the observer. The second requirement, on the other hand, imposes restrictions on the type of description by requiring that N be large.

Granted these conditions on the macroscopic description, we now argue as follows. If every part of the energy surface is accessible to phase points and if the time a phase point spends in a region A is, roughly speaking, proportional to the area of A, we can identify the region whose area is overwhelmingly the largest as the macroscopic equilibrium state, since it follows that:

1. If the system is not in the equilibrium state, it will almost always go into this state.
2. If the system is in the equilibrium state, it will almost always stay in this state, although fluctuations away from equilibrium to non-equilibrium states are almost certain to occur.

These ideas are due to Gibbs (similar ideas, although less precise and general, were put forward by Boltzmann) and can be made more precise by introducing probabilistic notions. Here we confine ourselves to a simple description of Gibbs's ideas on how a system might approach equilibrium.

First, we introduce the notion of probability by assuming that at time $t=0$ the system is in a definite macroscopic non-equilibrium state specified by the y_i variables, and is characterized by a probability density $\rho(p, q, 0)$ which is non-zero only in the given region of the energy surface corresponding to the given non-equilibrium state. The time development $\rho(p, q, t)$ is determined from Hamilton's equations and one would like $\rho(p, q, t)$ to approach the presumed equilibrium density $\rho(p, q)$ as t increases, independently of the initial density $\rho(p, q, 0)$.

Gibbs' description of how this might come about is as follows: As time evolves the initial region where $\rho(p, q, 0)$ is non-zero will change its shape drastically and be drawn out into a thin strip meandering over the energy surface until in a coarse-grained sense the distribution $\rho(p, q, t)$ becomes uniform on the energy surface. This is Gibbs's so-called *microcanonical distribution* formulated mathematically in the following section.

Although the above argument is extremely heuristic, it can be made more precise by appealing to the so-called ergodic theorems. It should be stressed, however, that the existence of a macroscopic description to begin with, which forms the basis of the argument, is not at all solved by ergodic theory. Ergodic theorems are necessary for the validity of Gibbs's argument, but they do not answer the basic question: How is a macroscopic description possible?

Readers interested in learning about ergodic theory and its relevance to statistical mechanics are referred to Uhlenbeck and Ford (1963) and Sinai (1976).

2.3 The Gibbs microcanonical distribution

The basic axiom, or postulate, of classical equilibrium statistical mechanics, due to Gibbs, is

Postulate I (Gibbs microcanonical distribution) *The equilibrium distribution of macroscopic states of an isolated conservative mechanical system is the uniform distribution on the energy surface.*

Thus, if our system has n degrees of freedom and is described by a Hamiltonian $\mathscr{H}=\mathscr{H}(p,q)$ with total (conserved) energy E, the microcanonical distribution is given by

$$\rho(p,q)=\delta(\mathscr{H}-E)/S(E) \tag{2.3.1}$$

where $\delta(x)$ is the Dirac delta function and

$$S(E)=\int\delta(\mathscr{H}-E)\mathrm{d}p\mathrm{d}q, \tag{2.3.2}$$

is the area of the energy surface $\mathscr{H}=E$ so that the distribution $\rho(p,q)$ is properly normalized. In eqn (2.3.2) the integration is over the entire $2n$-dimensional phase space and $\mathrm{d}p\mathrm{d}q$ denotes the Lebesgue measure for this space, which on occasion we will denote by $\mathrm{d}\Gamma$.

We will adopt the view here that eqn (2.3.1) is the basic axiom of classical equilibrium statistical mechanics and proceed immediately to discuss its consequences.

2.4 The canonical distribution

Since we are not usually concerned in the real world with isolated systems, we turn our attention now to systems coupled to a heat bath or reservoir. This enables us to introduce the notion of temperature and to make contact with the laws of thermodynamics.

Consider a system S that is 'weakly coupled' to, and in equilibrium with, a heat reservoir R. We assume that the combined system S $\cup$ R is isolated and that in equilibrium it is described by the microcanonical distribution. We now state

Postulate II (the Gibbs canonical distribution) *In the limit of an infinitely large heat reservoir* R, *the conditional probability density $D(p,q)$ for a system* S *in equilibrium with* R *is given by*

$$D(p,q)=Z^{-1}\exp(-\beta\mathscr{H}_{\mathrm{S}}) \tag{2.4.1}$$

where $\mathscr{H}_{\mathrm{S}}$ is the Hamiltonian of S, *$\beta=(kT)^{-1}$ with k Boltzmann's constant and T the absolute temperature of* R, *and Z is the canonical partition function defined by*

$$Z=\int\exp(-\beta\mathscr{H}_{\mathrm{S}})\mathrm{d}\Gamma_{\mathrm{s}} \tag{2.4.2}$$

where the integral in eqn (2.4.2) *is taken over the entire phase space of* S.

Unlike the microcanonical distribution, various 'specimen proofs' can be given of eqn (2.4.1) granted the microcanonical Postulate I. All such proofs, however, require a specific form for R, usually with a Hamiltonian of the form

$$\mathscr{H}_{\mathrm{R}}=\sum_{i=1}^{N}\mathscr{H}_i \tag{2.4.3}$$

where the N systems with Hamiltonians $\mathscr{H}_i$, $i=1, \ldots, N$, are weakly coupled to one another. In the absence of a general proof, eqn (2.4.1) may be considered as an independent postulate. To give some physical justification, however, we now present a proof due to Uhlenbeck and Ford (1963) for the special case where R is an ideal gas of N particles in a cube with Hamiltonian

$$\mathscr{H}_{\mathrm{R}}=\sum_{i=1}^{N}(\mathbf{p}_i)^2/2m. \tag{2.4.4}$$

Other proofs can be found in Grad (1952) and Khinchin (1949).

For simplicity, we take the Hamiltonian of the combined systems $\mathrm{S}\cup\mathrm{R}$ to be

$$\mathscr{H}=\mathscr{H}_{\mathrm{S}}+\mathscr{H}_{\mathrm{R}}. \tag{2.4.5}$$

Strictly speaking, we should include a coupling term $\mathscr{H}_{\mathrm{RS}}$ in eqn (2.4.5) so that the ergodic theorems are valid but since we can allow this term to be arbitrarily small, we avoid unnecessary complications by setting $\mathscr{H}_{\mathrm{RS}}$ equal to zero at the outset.

Assuming that the combined system is isolated, the microcanonical distribution for $\mathscr{H}$ is given by eqn (2.3.1),

$$\rho=\delta(\mathscr{H}_{\mathrm{S}}+\mathscr{H}_{\mathrm{R}}-E)/S(E) \tag{2.4.6}$$

where $S(E)$ is the area of the energy surface $\mathscr{H}_{\mathrm{S}}+\mathscr{H}_{\mathrm{R}}=E$. The conditional probability distribution D for the system S is obtained by integrating eqn (2.4.6) over the phase space of R. Using eqn (2.3.2), we thus obtain

$$\begin{aligned} D&=[S(E)]^{-1}\int\delta[\mathscr{H}_{\mathrm{R}}-(E-\mathscr{H}_{\mathrm{S}})]\mathrm{d}\Gamma_{\mathrm{R}} \\ &=S_{\mathrm{R}}(E-\mathscr{H}_{\mathrm{S}})/S(E) \end{aligned} \tag{2.4.7}$$

where S_{R} denotes the area of the reservoir energy surface.

Now since R is an ideal gas enclosed in a cube, of side L say, the energy surface $\mathscr{H}_{\mathrm{R}}=E$ of the reservoir from eqn (2.4.4) is a hypercylinder whose base is a hypersphere of radius $(2mE)^{1/2}$ in $3N$-dimensional momentum space, and whose orthogonal cross-section is a hypercube of side L in each of the $3N$ coordinate directions. It follows (see Problem 1) that

$$S_{\mathrm{R}}(E)=C_{N,L}E^{(3N-1)/2} \tag{2.4.8}$$

where we have indicated only the energy dependence.

We now write eqn (2.4.7) as

$$D=\frac{S_{\mathrm{R}}(E-\mathscr{H}_{\mathrm{S}})}{S_{\mathrm{R}}(E)}\left[\frac{S_{\mathrm{R}}(E)}{S(E)}\right] \tag{2.4.9}$$

and note from eqn (2.4.8) that

$$\lim_{N\to\infty} \frac{S_{\mathrm{R}}(E-\mathcal{H}_{\mathrm{S}})}{S_{\mathrm{R}}(E)} = \lim_{N\to\infty}\left(1-\frac{\mathcal{H}_{\mathrm{S}}}{E}\right)^{(3N-1)/2}$$

$$= \exp(-\beta\mathcal{H}_{\mathrm{S}}) \tag{2.4.10}$$

where

$$\beta = \lim_{N\to\infty}(3N/2E) \tag{2.4.11}$$

is assumed to be finite and, in deriving eqn (2.4.10), we have used the well-known result

$$\lim_{n\to\infty}\left(1-\frac{x}{n}\right)^{n} = \exp(-x). \tag{2.4.12}$$

Since D is a normalized distribution for all N, the required result, eqn (2.4.1), follows from eqns (2.4.9) and (2.4.10) where, from eqns (2.4.9), (2.4.10), and (2.4.2),

$$Z^{-1} = \lim_{N\to\infty}\frac{S_{\mathrm{R}}(E)}{S(E)} = [\int \exp(-\beta\mathcal{H}_{\mathrm{S}})\mathrm{d}\Gamma_{\mathrm{S}}]^{-1}. \tag{2.4.13}$$

It is to be noted from eqn (2.4.11) that β^{-1} is proportional to the (limiting) energy per particle of the heat reservoir. In this sense it is independent of the 'size' of R. Also, if we had several systems $\mathrm{S}_1, \mathrm{S}_2, \ldots, \mathrm{S}_n$ coupled to and in equilibrium with R, we would arrive at the same distribution eqn (2.4.1) with fixed β for each S_i. It follows that β must be a universal function of the absolute temperature T. In fact, for an ideal gas $E = 3NkT/2$, where k is Boltzmann's constant. We therefore take

$$\beta = (kT)^{-1} \tag{2.4.14}$$

in eqn (2.4.1).

One final point to note is that the above argument leading to eqn (2.4.1) does not depend on the size of S. The canonical distribution is therefore valid for even a single-particle system! On physical grounds, however, we would like the microcanonical and canonical distributions to give, where appropriate, equivalent thermodynamics. In particular, this means that in the canonical description the energy should be sharply peaked around its average value, which is intuitively the case for large systems only. We will return to this point later when we discuss the equivalence of the various distributions, or ensembles, as they are frequently called.

2.5 The laws of thermodynamics

Before we can proceed, we must set up a correspondence between statistical mechanical quantities and thermodynamic quantities and check that the laws of thermodynamics are satisfied.

Since we have presupposed equilibrium, the zeroth law is automatically satisfied. Also, since we are dealing at the outset with conservative mechanical systems, the total energy is conserved. The only problem with the *first law* then, is the distinction between *quantity of heat* and *external work.*

To produce changes of state we assume that we have external fields that are characterized by a set of parameters $a_1, \ldots, a_n$, which could, for example, represent successive components of a set of vectors. Included in the Hamiltonian then is a single-particle potential $U(\mathbf{r}_i; a_1, \ldots, a_n)$ for each particle labelled with index $i=1, 2, \ldots, N$, at position $\mathbf{r}_i$, due to the external fields. For example, we could have

$$U(\mathbf{r}_i; a_1, \ldots, a_{3n}) = \sum_{j=1}^{n} \phi_j(|\mathbf{r}_i - \mathbf{R}_j|) \tag{2.5.1}$$

which represents the potential produced by n fixed and independent centres of force at positions $\mathbf{R}_1, \ldots, \mathbf{R}_n$ whose successive components are $a_1, \ldots, a_{3n}$. In general, we assume that the Hamiltonian is a differentiable function of the a_is so that the 'force' in the direction of a_i is given by

$$F_i = -\frac{\partial \mathscr{H}}{\partial a_i}. \tag{2.5.2}$$

In statistical mechanics we calculate averages of certain quantities with respect to various probability distributions such as the microcanonical distribution (2.3.1) and the canonical distribution (2.4.1). Thus, the average of a phase function f is denoted by $\langle f \rangle$ and defined by

$$\langle f \rangle = \int f(p, q) D(p, q) \mathrm{d}\Gamma \tag{2.5.3}$$

where D could be, for example, the microcanonical or the canonical distribution.

From eqns (2.5.2) and (2.5.3) the average work done on a gas, for fixed position vectors $\mathbf{r}_i$ of the particles, by changing the a_is an infinitesimal amount δa_i, is given by

$$\delta W = -\sum_{i=1}^{n} \langle F_i \rangle \delta a_i. \tag{2.5.4}$$

For an isolated system, D is the microcanonical distribution ρ, and it is obvious in this case that $\delta W = \delta E$ (i.e. δ is adiabatic), where the average energy $E = \langle \mathscr{H} \rangle$ is just the total energy of the system.

For the canonical distribution, however, E, given from eqns (2.4.1) and (2.5.3) by

$$E = \langle \mathscr{H} \rangle = Z^{-1} \int \mathscr{H} \exp(-\beta \mathscr{H}) \mathrm{d}\Gamma, \tag{2.5.5}$$

can be changed by varying β (i.e. the temperature) while holding the a_is fixed. In this case it is clear that in general $\delta W \neq \delta E$, so we then *define*

$$\delta Q = \delta E - \delta W \tag{2.5.6}$$

to be the quantity of heat absorbed by the system due to the infinitesimal changes δa_i on the fields and $\delta\beta$ on the (inverse) temperature. This takes care of the first law.

To show that the *second law* is satisfied we must show (see Section 1.4) from the definitions (2.5.4), (2.5.5), and (2.5.6) that for an infinitesimal change δ in which both β and the a_i are changed reversibly while the system remains canonically distributed, $\beta\delta Q$ is an exact differential of the state of the system, specified now by β and the a_i.

To prove this we note first of all from eqn (2.5.5), and the definition eqn (2.4.2) of the canonical partition function Z, that the average energy is given by

$$E = -\frac{\partial}{\partial\beta}(\ln Z). \qquad (2.5.7)$$

It follows from this result, and from eqns (2.5.2) and (2.5.4), that

$$\delta E = -\frac{\partial^2}{\partial\beta^2}(\ln Z)\delta\beta - \sum_{i=1}^{n}\frac{\partial^2}{\partial\beta\partial a_i}(\ln Z)\delta a_i \qquad (2.5.8)$$

and

$$\delta W = Z^{-1}\int\left(\sum_{i=1}^{n}\frac{\partial\mathscr{H}}{\partial a_i}\delta a_i\right)\exp(-\beta\mathscr{H})\mathrm{d}\Gamma$$

$$= -\beta^{-1}\sum_{i=1}^{n}\frac{\partial}{\partial a_i}(\ln Z)\delta a_i. \qquad (2.5.9)$$

It is now a straightforward matter to show from eqns (2.5.6), (2.5.8), and (2.5.9) (see Problem 4) that

$$\beta\delta Q = \beta(\delta E - \delta W) = \delta\left[-\beta^2\frac{\partial}{\partial\beta}(\beta^{-1}\ln Z)\right], \qquad (2.5.10)$$

which implies the required result, that $\beta\delta Q$ is an exact differential of a state function.

Comparison with the thermodynamic formula (1.4.16), $\mathrm{d}S = \delta Q/T$, noting that $\beta = (kT)^{-1}$, shows that the entropy S is given in terms of the canonical partition function Z by

$$S = \frac{\partial}{\partial T}(kT\ln Z) + \text{constant}. \qquad (2.5.11)$$

It is clear that the additive constant in eqn (2.5.11) cannot be determined from the arguments presented here, so in a sense the constant is arbitrary. Its dependence on the number N of particles in the system, for example, can only be agreed upon by convention. The normal convention requires that the entropy be an *extensive quantity*, which means that for fixed T it should be asymptotically proportional to the size of the system. More precisely, for simple one-component systems of N particles confined to a region with volume V, we require the limit

$$\lim_{\substack{N, V\to\infty \\ v=V/N \text{ fixed}}} N^{-1}S(V,N,T)=s(v,T) \tag{2.5.12}$$

to exist and depend only on the *intensive quantities* (i.e. independent of the size of the system) T and '*specific volume*' v.

The limit in eqn (2.5.12) is called the *thermodynamic limit* or *bulk limit* and will be discussed in some detail in Chapter 3. Suffice it to say here that the additive constant in eqn (2.5.11) can be interpreted as a multiplicative constant for Z which is usually taken to be $(N!)^{-1}$. That is, in order to obtain *extensive thermodynamics* we redefine the canonical partition function by

$$Z=(N!)^{-1}\int\exp(-\beta\mathscr{H})\,\mathrm{d}\Gamma. \tag{2.5.13}$$

In the following section, we develop further connections between statistical mechanical quantities and thermodynamic quantities discussed in Chapter 1.

2.6 Derived thermodynamic quantities for fluid and magnetic systems

If one compares eqn (2.5.11) for the entropy with eqn (1.5.11) it is clear that an obvious candidate for a statistical mechanical definition of the Helmholtz free energy $\Psi(V, N, T)$ for a fluid system of N particles confined to a region with volume V and described microscopically by a Hamiltonian $\mathscr{H}$, is

$$\Psi(V,N,T)=-kT\ln Z(V,N,T) \tag{2.6.1}$$

where $Z(V,N,T)$ is the canonical partition function eqn (2.5.13). Indeed, if one takes note of the expression given in eqn (2.5.7) for the average energy $E=\langle\mathscr{H}\rangle$, one obtains, using the fact that $\beta=(kT)^{-1}$, the relation

$$E=\Psi-T\left(\frac{\partial\Psi}{\partial T}\right)_V. \tag{2.6.2}$$

The coefficient of T in eqn (2.6.2) is simply $-S$ from eqn (2.5.11) so that if one interprets E as the internal energy U, eqn (2.6.2) conforms exactly to the thermodynamic definition

$$\Psi=U-TS \tag{2.6.3}$$

given in eqn (1.5.1).

Having identified the Helmholtz free energy, other thermodynamic quantities can be obtained straightforwardly by differentiation with respect to appropriate variables. For example, the pressure P is given by

$$P=-\left(\frac{\partial\Psi}{\partial V}\right)_T, \tag{2.6.4}$$

the specific heat at constant volume by

$$C_V = \left(\frac{\partial E}{\partial T}\right)_V = -T\left(\frac{\partial^2 \Psi}{\partial T^2}\right)_V, \tag{2.6.5}$$

and so forth.

It will be observed from the above that calculations using the canonical distribution essentially begin and end with an evaluation of the partition function Z from eqn (2.5.13).

For magnetic systems the microscopic objects are 'spins', rather than particles, which may be thought of as vectors or scalars occupying fixed points in space. If we have a set of N such spins $\Sigma = \{\mathbf{S}_i | \; i = 1, 2, \ldots N\}$, satisfying some condition, such as having unit length, forming our phase space or *configuration space*, with some (internal) interaction energy $U(\Sigma)$, the Hamiltonian for such a system in the presence of a fixed external magnetic field $\mathbf{H}$ is given by

$$\mathscr{H}(\Sigma) = U(\Sigma) - \mathbf{H} \cdot \mathbf{M}(\Sigma) \tag{2.6.6}$$

where

$$\mathbf{M}(\Sigma) = \sum_{i=1}^{N} \mathbf{S}_i \tag{2.6.7}$$

is the magnetization of the system in configuration Σ. By analogy with eqn (2.5.13) we define the canonical partition function for such a system by

$$Z(H, N, T) = C_N \int \exp(-\beta \mathscr{H}(\Sigma))\, d\Sigma \tag{2.6.8}$$

where the integral is taken over the configuration space of the spins and C_N is some constant which is chosen to make the thermodynamics extensive.

If we take for simplicity a situation in which the external magnetic field and the spins are scalar quantities, the average magnetization is given from eqns (2.6.6)–(2.6.8) by

$$\begin{aligned} M &= \int M(\Sigma) \exp(-\beta \mathscr{H})\, d\Sigma \Big/ \int \exp(-\beta \mathscr{H})\, d\Sigma \\ &= \frac{\partial}{\partial H}(kT \ln Z(H, N, T))_T. \end{aligned} \tag{2.6.9}$$

The entropy, by analogy with eqn (2.5.11), is given by

$$S(H, N, T) = \frac{\partial}{\partial T}(kT \ln Z(H, N, T))_H. \tag{2.6.10}$$

Comparison with the corresponding formulae (1.5.17) and (1.5.18) then leads us to define the *Gibbs free energy* $\Phi(H, N, T)$ by

$$\Phi(H, N, T) = -kT \ln Z(H, N, T). \tag{2.6.11}$$

From eqns (2.6.6) and (2.6.11) the average energy is given by

$$\langle \mathscr{H} \rangle = \langle U \rangle - HM$$

$$= \Phi - T\left(\frac{\partial \Phi}{\partial T}\right)_H, \tag{2.6.12}$$

which, on rearranging, conforms to the thermodynamic definition given in eqn (1.5.3) if $\langle U \rangle$ in eqn (2.6.12) is interpreted as the internal energy of the system. Just as for fluid systems, further thermodynamic quantities can now be obtained by taking appropriate derivatives of Φ in eqn (2.6.11). For example, the isothermal susceptibility is given by

$$\chi_T = \left(\frac{\partial M}{\partial H}\right)_T = -\left(\frac{\partial^2 \Phi}{\partial H^2}\right)_T \tag{2.6.13}$$

and the specific heat at constant field by

$$C_H = T\left(\frac{\partial S}{\partial T}\right)_H = -T\left(\frac{\partial^2 \Phi}{\partial T^2}\right)_H. \tag{2.6.14}$$

Ultimately we will be interested in the *thermodynamic limit* or bulk limit. For fluid systems we take $N, V \to \infty$ with fixed specific volume $v = V/N$ and define

$$\psi(v, T) = \lim_{\substack{N, V \to \infty \\ v = V/N \text{ fixed}}} N^{-1} \Psi(V, N, T) \tag{2.6.15}$$

when it exists, to be the limiting free energy per particle. Similarly, for magnetic systems, we define the limiting free energy per spin by

$$\psi(H, T) = \lim_{N \to \infty} N^{-1} \Phi(H, N, T). \tag{2.6.16}$$

In the sequel we will refer to ψ, defined either by eqn (2.6.15) or (2.6.16), as *the* free energy; it being understood that in the case of a fluid we are dealing with the Helmholtz free energy and in the magnetic case with the Gibbs free energy.

Whenever the thermodynamic limit exists we refer to the system as having the *extensive property*, which simply means that thermodynamic quantities such as entropy, magnetization, specific heat, etc. are (asymptotically) proportional to the size of the system. The problem here is to find conditions on the microscopic interactions (or Hamiltonian) that will guarantee the existence of the thermodynamic limit. (This question is relevant to the discussion which will follow on the equivalence of various statistical mechanical descriptions. As we will see later, it is of fundamental importance in a rigorous discussion of phase transitions.) The existence of the thermodynamic limit will be discussed in some detail in Chapter 3.

2.7 The grand-canonical distribution

We began our discussion of statistical mechanics with the microcanonical distribution which applies to isolated systems with a fixed number of particles N in a region having fixed volume V and having a fixed total energy E. We then considered more realistic systems with fixed N and V but with variable energy E in contact with a heat reservoir at fixed temperature T. Such systems are described by the canonical distribution. In many situations we find that N varies as well as E while V and T are fixed. For example, we could have chemical reactions taking place in an open vessel, with fixed V and T, in contact with a 'particle reservoir'. For such systems we have the so-called 'grand-canonical distribution'.

For canonically distributed systems we saw that thermodynamic quantities of physical interest for a system described by a Hamiltonian $\mathscr{H}$ can be obtained from the canonical partition function

$$Z(V,N,T)=(N!)^{-1}\int\exp(-\beta\mathscr{H})\,\mathrm{d}\Gamma. \tag{2.7.1}$$

Similarly, quantities of interest for open systems described by the grand-canonical distribution can be obtained from the *grand-canonical partition function* Z_{G} defined by

$$Z_{\mathrm{G}}(V,T,z)=\sum_{N=0}^{\infty} z^N Z(V,N,T) \tag{2.7.2}$$

where the auxiliary variable z, called the fugacity or activity, is related to the *chemical potential* of the system μ by (see Problem 9)

$$z=\exp(\beta\mu). \tag{2.7.3}$$

The (normalized) grand-canonical distribution function ρ for the number of particles N is given by

$$\rho(N)=[Z_{\mathrm{G}}(V,T,z)]^{-1}z^N Z(V,N,T) \tag{2.7.4}$$

so that the average number of particles is given by

$$\begin{aligned}\langle N\rangle &= \sum_{N=0}^{\infty} N\rho(N)\\ &= z\frac{\partial}{\partial z}\ln Z_{\mathrm{G}}(V,T,z).\end{aligned} \tag{2.7.5}$$

The grand-canonical pressure p is defined by

$$p=kTV^{-1}\ln Z_{\mathrm{G}}(V,T,z) \tag{2.7.6}$$

and the equation of state is obtained, from eqns (2.7.5) and (2.7.6), by eliminating the fugacity z.

In order to match up this definition of pressure with, say, the canonical definition we must proceed to the thermodynamic limit. As we will see in the following section, it is only in this limit that the microcanonical, canonical, and grand-canonical distributions give identical thermodynamic descriptions.

For large V it is clear that the right-hand side of eqn (2.7.2) will be dominated by the maximum term in the sum, corresponding to say $N=N_0$ which we will assume is asymptotically proportional to V for large V. Using Laplace's method and assuming that

$$\lim_{\substack{V\to\infty \\ v=V/N_0 \text{ fixed}}} N_0^{-1} \ln Z(V, N_0, T) = -\beta\psi(v, T) \tag{2.7.7}$$

exists, we have that

$$\begin{aligned} \chi(T, z) &= \lim_{V\to\infty} V^{-1} \ln Z_G(V, T, z) \\ &= v^{-1}(\ln z - \beta\psi(v, T)). \end{aligned} \tag{2.7.8}$$

The function $\chi(T, z)$ is commonly called the grand-canonical potential. Differentiating eqn (2.7.8) with respect to z, we obtain

$$z\frac{\partial \chi}{\partial z} = v^{-1} \tag{2.7.9}$$

which agrees with eqn (2.7.5) in the limit $V\to\infty$ if we identify the maximizing $N_0 = v^{-1}V$ in eqn (2.7.2) with the average number of particles $\langle N \rangle$ in eqn (2.7.5). 'Solving' eqn (2.7.9) for z in terms of v and substituting into eqn (2.7.8) gives the relation

$$\psi(v, T) = kT \ln z(v) - kTv\chi(T, z(v)) \tag{2.7.10}$$

between the canonical free energy ψ and the grand-canonical potential χ.

Finally, if we differentiate eqn (2.7.10) with respect to v and use the canonical definition of pressure p (per particle) we obtain, using eqns (2.7.8) and (2.7.9), the expression

$$\begin{aligned} p &= -\left(\frac{\partial \psi}{\partial v}\right)_T \\ &= kT\left(\frac{1}{z} - v\frac{\partial \chi}{\partial z}\right)\frac{\partial z}{\partial v} + kT\chi(T, z(v)) \\ &= kT\chi(T, z(v)) \end{aligned} \tag{2.7.11}$$

which is identical with the grand-canonical definition (2.7.6) in the limit $V\to\infty$.

A more rigorous discussion of the above manipulations, particularly the existence of the limit in eqn (2.7.8) and the solubility of eqn (2.7.9) can be found in the original work of Yang and Lee (1952).

2.8 Equivalence of the microcanonical, canonical, and grand-canonical descriptions

The manipulations in the preceding section leading to eqn (2.7.9) and the resulting identification of the maximizing N_0 in eqn (2.7.2) with the mean or average number $\langle N\rangle$ essentially means that, for large volume V, the distribution of the number of particles in the grand-canonical distribution is very sharply peaked around $\langle N\rangle$. To make this more precise we consider the root-mean-square deviation of the number of particles N from its mean value, defined by

$$\Delta N = \left\{\frac{\langle (N-\langle N\rangle)^2\rangle}{\langle N\rangle^2}\right\}^{1/2} \tag{2.8.1}$$

If we set

$$\langle N\rangle = v^{-1}V \tag{2.8.2}$$

and use eqn (2.7.5) to define $v=v(z)$ by

$$[v(z)]^{-1} = z\frac{\partial}{\partial z}(V^{-1}\ln Z_G(V,T,z)), \tag{2.8.3}$$

it is not difficult to show (see Problem 11) that for fixed V,

$$(\Delta N)^2 = -V^{-1}z\frac{\partial v}{\partial z}. \tag{2.8.4}$$

On the other hand, from (2.8.3) and the definition (2.7.6) of pressure in the grand-canonical description, we have

$$z\frac{\partial p}{\partial z} = kTv^{-1}. \tag{2.8.5}$$

It then follows, from eqn (2.8.4), that

$$\begin{aligned}(\Delta N)^2 &= -V^{-1}\left(\frac{\partial v}{\partial p}\right)\left(z\frac{\partial p}{\partial z}\right)\\ &= V^{-1}kTK_T \end{aligned} \tag{2.8.6}$$

where

$$K_T = -\left(v\frac{\partial p}{\partial v}\right)^{-1} \tag{2.8.7}$$

is the isothermal compressibility per article.

So long as we are away from a critical point, K_T remains finite as $V\to\infty$. It then follows from eqn (2.8.6) that, except possibly at a critical point, the fluctuation ΔN in particle number, or density, for fixed V vanishes as $V^{-1/2}$ in the thermodynamic limit. At a critical point where K_T may be expected to diverge as V^{σ} when $V\to\infty$, the fluctuation ΔN will still vanish in the thermodynamic limit, provided $0<\sigma<1$. Various heuristic scaling arguments suggest that this is indeed the case, so that while comparatively large fluctuations in density may occur at a critical point, it is expected that such behaviour will not invalidate the equivalence of the canonical and grand-canonical descriptions.

Similarly, to show that the microcanonical and canonical distributions give identical thermodynamic behaviour in the thermodynamic limit $N, V\to\infty$ with $v=V/N$ fixed, we must show that fluctuations in energy approach zero in this limit.

Thus, it is not difficult to show that the root-mean-square deviation of the energy for a finite system with Hamiltonian $\mathscr{H}$ is given by (see Problem 7)

$$\Delta E=\left\{\frac{\langle(\mathscr{H}-\langle\mathscr{H}\rangle)^2\rangle}{\langle\mathscr{H}\rangle^2}\right\}^{1/2}=\left\{\frac{kT^2C_V}{\langle\mathscr{H}\rangle^2}\right\}^{1/2} \tag{2.8.8}$$

where the angular brackets now denote canonical average and C_V is the (extensive) specific heat at constant volume.

Since the average energy $\langle\mathscr{H}\rangle$ and C_V are both extensive quantities, i.e. proportional to the size of the system, it follows that fluctuations of energy vanish in the thermodynamic limit, except possibly at a phase transition point where the specific heat (per particle) may diverge. Again, however, we expect on the basis of various heuristic arguments, that at a critical point the rate of divergence of the specific heat per particle will be a power σ of V smaller than unity and hence that the microcanonical and canonical ensembles give equivalent descriptions for all temperatures and densities.

2.9 Problems

1. Show that the surface area of an n-dimensional sphere of radius r is given by

$$S_n(r)=2\pi^{n/2}r^{n-1}/\Gamma(n/2)$$

where $\Gamma(\alpha)$ is the gamma function.

2. The entropy $S(V,N,E)$ for an isolated system of N particles with total energy E confined to a region having volume V is defined by

$$S(V,N,E)=k\ln[S(E)/N!]$$

where $S(E)$ is the area of the energy surface $\mathscr{H}=E$.

Using the result of Problem 1 and Stirling's formula $\Gamma(n)\sim n^n e^{-n}$ as $n\to\infty$, show for an ideal gas with Hamiltonian

$$\mathscr{H}=\sum_{i=1}^{N}(\mathbf{p}_i)^2/2m$$

that

$$S(V,N,E)\sim Nk\left\{\ln\left[v\left(\frac{4\pi mE}{3N}\right)^{3/2}\right]+5/2\right\}\qquad \text{as } N\to\infty$$

where $v=V/N$ is the specific volume.

3. (a) If $S(V, E)$ is the entropy of an isolated system show that

$$dS = \left(\frac{\partial S}{\partial E}\right)_V dE + \left(\frac{\partial S}{\partial V}\right)_E dV$$

is equivalent to the first law of thermodynamics if we define 'temperature' T and 'pressure' P for such a system by

$$T = \left(\frac{\partial S}{\partial E}\right)_V^{-1} \quad \text{and} \quad P = T\left(\frac{\partial S}{\partial V}\right)_E.$$

(b) Use the definitions of part (a) and the results of Problem 2 to show that in the microcanonical description, the equation of state for an ideal gas is given by

$$PV = NkT$$

and that

$$E = \tfrac{3}{2}NkT.$$

4. If Z is the canonical partition function and

$$\delta E = -\frac{\partial^2}{\partial \beta^2}(\ln Z)\,\delta\beta - \sum_{i=1}^{n} \frac{\partial^2}{\partial\beta\partial a_i}(\ln Z)\,\delta a_i$$

and

$$\delta W = -\beta^{-1} \sum_{i=1}^{n} \frac{\partial}{\partial a_i}(\ln Z)\,\delta a_i$$

are the changes in internal energy and work done on a system resulting from changes in temperature and external field parameters a_i, show that $\beta(\delta E - \delta W)$ is an exact differential of the function (cf. eqns (2.5.8)–(2.5.10))

$$-\beta^2 \frac{\partial}{\partial \beta}(\beta^{-1} \ln Z) + \text{constant}.$$

5. Show that the canonical partition function, eqn (2.5.13), for an ideal gas of N particles contained in a region with volume V is given by

$$Z(V, N, T) = \frac{V^N}{N!}(2\pi mkT)^{3N/2}.$$

Deduce that the equation of state and the expression for the average energy are identical with the corresponding microcanonical expressions given in Problem 3(b).

6. For a system of N particles contained in a region with volume V and Hamiltonian

$$\mathscr{H} = \sum_{i=1}^{N} \frac{\mathbf{p}_i^2}{2m} + U(\mathbf{r}_1, \ldots, \mathbf{r}_N),$$

where $\mathbf{r}_i$ and $\mathbf{p}_i$ are the position and momentum of the ith particle, the canonical partition function is given by

$$Z(V,N,T)=\frac{1}{N!}\int \exp(-\beta\mathscr{H})\,\mathrm{d}r\mathrm{d}p$$

$$=\frac{(2\pi mkT)^{3N/2}}{N!}I(V,T)$$

where

$$I(V,T)=\int_V \cdots \int \exp(-\beta U(\mathbf{r}_1,\ldots,\mathbf{r}_N))\,\mathrm{d}\mathbf{r}_1\cdots\mathrm{d}\mathbf{r}_N.$$

Assuming that the potential energy U is a homogeneous function of degree n, i.e.

$$U(\lambda\mathbf{r}_1,\lambda\mathbf{r}_2,\ldots,\lambda\mathbf{r}_N)=\lambda^n U(\mathbf{r}_1,\ldots,\mathbf{r}_N)$$

for all real λ, show that $V^{-N}I(V,T)$ is a function of the single variable TV^{-n}.

7. If Z is the canonical partition function and an average $\langle A\rangle$ of a quantity A of the state of the system is defined by

$$\langle A\rangle=Z^{-1}\int A\exp(-\beta\mathscr{H})\,\mathrm{d}\Gamma$$

where $\mathscr{H}$ is the Hamiltonian of the system, show that

$$\langle\mathscr{H}^2\rangle-\langle\mathscr{H}\rangle^2=kT^2C_V$$

and deduce from this result that the specific heat C_V at constant volume is non-negative.

8. The Hamiltonian for a set of N non-interacting scalar spins μ_i, $i=1,2,\ldots,N$ in the presence of an external magnetic field H is given by

$$\mathscr{H}\{\mu\}=-H\sum_{i=1}^{N}\mu_i$$

and the canonical partition function is defined (in general) by

$$Z(H,N,T)=\sum_{\{\mu\}}\exp(-\beta\mathscr{H}\{\mu\})$$

where the sum is taken over all allowed configurations $\{\mu\}=(\mu_1,\mu_2,\ldots,\mu_N)$. In the particular case where spins take only two values $\mu_i=+1$ (spin up) and $\mu_i=-1$ (spin down) show that the average magnetization is given by

$$M=\left\langle\sum_{i=1}^{N}\mu_i\right\rangle=N\tanh\beta H.$$

A system such as this is called a *paramagnet*.

9. In the grand-canonical description the average number of particles N is given in terms of the grand-canonical partition function Z_G by

$$N=z\frac{\partial}{\partial z}\ln Z_G(V,T,z).$$

If we define

$$\Psi(N,V,T)=NkT\ln z-kT\ln Z_G(V,T,z)$$

with $z=z(N, V, T)$ determined from the previous equation, show that the chemical potential μ is given by

$$\mu \equiv \left(\frac{\partial \Psi}{\partial N}\right)_{V,T} = kT \ln z.$$

10. Show that the grand-canonical partition function for a system confined to a region with volume V and described by the Hamiltonian

$$\mathscr{H} = \sum_{i=1}^{N} \left[\frac{\mathbf{p}_i^2}{2m} + u(\mathbf{r}_i)\right]$$

is given by

$$Z_G(V, T, z) = \exp(zq(V, T))$$

where

$$q(V, T) = (2\pi mkT)^{3/2} \int_V \exp(-\beta u(\mathbf{r}))\,\mathrm{d}\mathbf{r}.$$

Deduce that the equation of state in the grand-canonical description of this system is always of the ideal-gas form regardless of the single-particle potential $u(\mathbf{r})$.

11. Show that the mean-square fluctuation of the number of particles in the grand-canonical description can be expressed in the form

$$\langle N^2 \rangle - \langle N \rangle^2 = z\frac{\partial}{\partial z} z \frac{\partial}{\partial z} \ln Z_G(V, T, z).$$

Deduce that for an ideal gas the root-mean-square fluctuation of N is given by

$$\left\{\frac{\langle (N - \langle N \rangle)^2 \rangle}{\langle N \rangle^2}\right\}^{1/2} = \langle N \rangle^{-1/2}.$$

2.10 Solutions to problems

1. If $S_n(r)$ is the surface area of the n-dimensional sphere

$$\sum_{i=1}^{n} x_i^2 = r^2,$$

we have

$$\int_0^\infty \mathrm{e}^{-pr^2} S_n(r)\,\mathrm{d}r = \int_{-\infty}^{\infty} \cdots \int \exp\left(-p \sum_{i=1}^{n} x_i^2\right) \mathrm{d}x_1 \cdots \mathrm{d}x_n$$
$$= (\pi/p)^{n/2}.$$

The stated result follows by changing variables to $x = r^2$ and noting that the Laplace transform of x^{k-1} is given by

$$\int_0^\infty \mathrm{e}^{-px} x^{k-1}\,\mathrm{d}x = p^{-k}\Gamma(k).$$

2. The energy surface is a hypercylinder whose base is a hypersphere of radius $(2mE)^{1/2}$ in $3N$-dimensional momentum space and whose orthogonal cross-section is a hypercube of volume V for each of the N spatial coordinates. It follows from Problem 1 that

$$\begin{aligned} S(E)/N! &= V^N 2\pi^{3N/2}(2mE)^{(3N-1)/2}/\Gamma(3N/2)\,N! \\ &\sim (2\pi mE)^{3N/2}(V/N)^N/(3N/2)^{3N/2}\exp(-5N/2) \\ &= \{[4\pi mE/3N]^{3/2} v \exp(5/2)\}^N \end{aligned}$$

which gives the stated result.

3. (a) The first law of thermodynamics can be expressed in the form

$$\delta Q = \mathrm{d}E + P\mathrm{d}V.$$

By definition

$$\delta Q = T\mathrm{d}S = T\left(\frac{\partial S}{\partial E}\right)_V \mathrm{d}E + T\left(\frac{\partial S}{\partial V}\right)_E \mathrm{d}V.$$

The result follows by comparing these two expressions for δQ.

(b) From Problem 2

$$\left(\frac{\partial S}{\partial E}\right)_V = 3Nk/2E \quad \text{and} \quad \left(\frac{\partial S}{\partial V}\right)_E = Nk/V.$$

It then follows, from part (a), that

$$P = T\left(\frac{\partial S}{\partial V}\right)_E = NkT/V$$

and

$$T = \left(\frac{\partial S}{\partial E}\right)_V^{-1} = 2E/3Nk$$

which give the required results.

4.
$$\begin{aligned} &\delta\left[-\beta^2 \frac{\partial}{\partial\beta}(\beta^{-1}\ln Z) + \text{constant}\right] \\ &= \delta\left(\ln Z - \beta\frac{\partial}{\partial\beta}\ln Z\right) \\ &= \frac{\partial}{\partial\beta}(\ln Z)\,\delta\beta + \sum_{i=1}^{n}\frac{\partial}{\partial a_i}(\ln Z)\,\delta a_i \\ &\quad -\delta\beta\frac{\partial}{\partial\beta}(\ln Z) - \beta\frac{\partial^2}{\partial\beta^2}(\ln Z)\,\delta\beta - \beta\sum_{i=1}^{n}\frac{\partial^2}{\partial\beta\partial a_i}(\ln Z)\,\delta a_i \\ &= \beta(\delta E - \delta W). \end{aligned}$$

5. The canonical partition function for an ideal gas is given by

$$Z(V,N,T)=(N!)^{-1}\int_V\cdots\int \mathrm{d}\mathbf{r}_1\cdots\mathrm{d}\mathbf{r}_N\int_{-\infty}^{\infty}\cdots\int \mathrm{d}\mathbf{p}_1\cdots\mathrm{d}\mathbf{p}_N$$

$$\times\exp\left(-\beta\sum_{i=1}^{N}\mathbf{p}_i^2/2m\right)$$

$$=(N!)^{-1}V^N\left\{\int_{-\infty}^{\infty}\exp(-\beta p^2/2m)\,\mathrm{d}p\right\}^{3N}$$

$$=(N!)^{-1}V^N(2\pi m/\beta)^{3N/2}.$$

The pressure is given by

$$P=\frac{\partial}{\partial V}[kT\ln Z(V,N,T)]=NkT/V$$

and the average energy is given by

$$E=-\frac{\partial}{\partial\beta}(\ln Z)=3N/2\beta=3NkT/2$$

which agree with the microcanonical expressions in Problem 3.

6. If we make the substitution $\mathbf{r}_i\to\lambda\mathbf{r}_i$ in the integral defining $I(V,T)$ we obtain

$$I(V,T)=\lambda^N\int_{\lambda^{-1}V}\cdots\int\exp[-\beta U(\lambda\mathbf{r}_1,\ldots,\lambda\mathbf{r}_N)]\,\mathrm{d}\mathbf{r}_1\cdots\mathrm{d}\mathbf{r}_N$$

$$=\lambda^N\int_{\lambda^{-1}V}\cdots\int\exp[-\beta\lambda^n U(\mathbf{r}_1,\ldots,\mathbf{r}_N)]\,\mathrm{d}\mathbf{r}_1\cdots\mathrm{d}\mathbf{r}_N$$

$$=\lambda^N I(\lambda^{-1}V,\lambda^{-n}T)$$

which is valid for arbitrary real positive λ. In particular, if we choose $\lambda=V$, we obtain

$$I(V,T)=V^N I(1,V^{-n}T)$$

which gives the required result.

7. By definition

$$C_V=\frac{\partial}{\partial T}\langle\mathscr{H}\rangle=-\frac{1}{kT^2}\frac{\partial}{\partial\beta}\langle\mathscr{H}\rangle$$

$$=-\frac{1}{kT^2}\frac{\partial}{\partial\beta}\left[Z^{-1}\int\mathscr{H}\exp(-\beta\mathscr{H})\,\mathrm{d}\Gamma\right]$$

$$=\frac{1}{kT^2}\left\{Z^{-2}\frac{\partial Z}{\partial\beta}\int\mathscr{H}\exp(-\beta\mathscr{H})\,\mathrm{d}\Gamma+Z^{-1}\int\mathscr{H}^2\exp(-\beta\mathscr{H})\,\mathrm{d}\Gamma\right\}$$

$$=\frac{1}{kT^2}(-\langle\mathscr{H}\rangle^2+\langle\mathscr{H}^2\rangle)$$

where in the last step we have used the fact that

$$\frac{\partial Z}{\partial\beta}=\frac{\partial}{\partial\beta}\int\exp(-\beta\mathscr{H})\,\mathrm{d}\Gamma=-\int\mathscr{H}\exp(-\beta\mathscr{H})\,\mathrm{d}\Gamma=-Z\langle\mathscr{H}\rangle.$$

C_V non-negative follows from the elementary identity

$$\langle (A-\langle A\rangle)^2\rangle = \langle A^2\rangle - \langle A\rangle^2.$$

8. The canonical partition function is given by

$$\begin{aligned} Z(H,N,T) &= \sum_{\{\mu\}} \exp\left(\beta H \sum_{i=1}^{N} \mu_i\right) \\ &= \left[\sum_{\mu=\pm 1} \exp(\beta H \mu)\right]^N \\ &= (2\cosh \beta H)^N. \end{aligned}$$

The required result follows by noting that

$$\begin{aligned} \left\langle \sum_{i=1}^{N} \mu_i \right\rangle &= Z^{-1} \sum_{\{\mu\}} \left(\sum_{i=1}^{N} \mu_i\right) \exp(-\beta \mathscr{H}\{\mu\}) \\ &= \frac{\partial}{\partial \beta H} (\ln Z). \end{aligned}$$

9. Considering z, the solution of

$$N = z \frac{\partial}{\partial z} \ln Z_G(V,T,z),$$

to be a function of N, V, and T, we have

$$\begin{aligned} \mu &= \frac{\partial}{\partial N} [NkT \ln z - kT \ln Z_G(V,T,z)] \\ &= kT \ln z + \left[NkTz^{-1} - kT \frac{\partial}{\partial z} \ln Z_G(V,T,z)\right] \left(\frac{\partial z}{\partial N}\right)_{V,T} \\ &= kT \ln z. \end{aligned}$$

10. The canonical partition function is given by

$$\begin{aligned} Z(V,N,T) &= \frac{1}{N!} \int_V \cdots \int \exp\left[-\beta \sum_{i=1}^{N} u(\mathbf{r}_i)\right] d\mathbf{r}_1 \cdots d\mathbf{r}_N \\ &\quad \times \int_{-\infty}^{\infty} \cdots \int \exp\left(-\beta \sum_{i=1}^{N} \mathbf{p}_i^2/2m\right) d\mathbf{p}_1 \cdots d\mathbf{p}_N \\ &= \frac{1}{N!} \left\{\int_V \exp[-\beta u(\mathbf{r})]\, d\mathbf{r}\right\}^N \left\{\int_{-\infty}^{\infty} \exp(-\beta p^2/2m)\, dp\right\}^{3N} \\ &= [q(V,T)]^N/N! \end{aligned}$$

where

$$q(V,T) = (2\pi mkT)^{3/2} \int_V \exp[-\beta u(\mathbf{r})]\, d\mathbf{r}.$$

The grand-canonical partition function is then given by

$$Z_G(V,T,z) = \sum_{N=0}^{\infty} z^N Z(V,N,T) = \exp[zq(V,T)].$$

The pressure P and average number of particles N are then given by

$$PV = kT \ln Z_G(V, T, z) = kTzq(V, T)$$

and

$$N = z \frac{\partial}{\partial z} \ln Z_G(V, T, z) = zq(V, T).$$

Elimination of z between these two equations gives the ideal gas equation of state

$$PV = NkT.$$

11. In the grand-canonical description the average number of particles is given by

$$\begin{aligned}\langle N \rangle &= Z_G^{-1} \sum_{N=0}^{\infty} N z^N Z(V, N, T) \\ &= z \frac{\partial}{\partial z} (\ln Z_G(V, T, z)).\end{aligned}$$

Differentiating this expression with respect to z gives

$$\begin{aligned}& z \frac{\partial}{\partial z}\left(z \frac{\partial}{\partial z} (\ln Z_G(V, T, z))\right) \\ &= Z_G^{-1} \sum_{N=0}^{\infty} N^2 z^N Z(V, N, T) - Z_G^{-2} z \frac{\partial Z_G}{\partial z} \sum_{N=0}^{\infty} N z^N Z(V, N, T) \\ &= \langle N^2 \rangle - z \frac{\partial}{\partial z} (\ln Z_G) Z_G^{-1} \sum_{N=0}^{\infty} N z^N Z(V, N, T) \\ &= \langle N^2 \rangle - \langle N \rangle^2.\end{aligned}$$

For the ideal gas (see Problem 10),

$$Z_G(V, T, z) = \exp [zq(V, T)]$$

where

$$q(V, T) = V(2\pi mkT)^{3/2}.$$

From the above formula we then obtain

$$\begin{aligned}\langle N^2 \rangle - \langle N \rangle^2 &= \langle (N - \langle N \rangle)^2 \rangle \\ &= z \frac{\partial}{\partial z} z \frac{\partial}{\partial z} \ln Z_G(V, T, z) \\ &= zq(V, T) \\ &= z \frac{\partial}{\partial z} \ln Z_G(V, T, z) \\ &= \langle N \rangle\end{aligned}$$

which gives the required result.

3

MODEL SYSTEMS AND THE THERMODYNAMIC LIMIT

3.1 Introduction

As we saw in Chapter 2, calculations in equilibrium statistical mechanics begin and essentially end with an evaluation of the canonical partition function defined by

$$Z(V, N, T) = \frac{1}{N!} \int \exp(-\beta \mathscr{H}) \mathrm{d}p \mathrm{d}q \tag{3.1.1}$$

where $\mathscr{H}$ is the microscopic Hamiltonian of a system of N particles confined to a region with volume V, $\beta = (kT)^{-1}$ where k is Boltzmann's constant and T the absolute temperature, and the integration in eqn (3.1.1) is taken over the $6N$-dimensional phase space.

Here and henceforth we will consider only systems with Hamiltonians of the form

$$\mathscr{H} = \sum_{i=1}^{N} \frac{\mathbf{p}_i^2}{2m} + \sum_{1 \leqslant i < j \leqslant N} \phi(|\mathbf{r}_i - \mathbf{r}_j|) \tag{3.1.2}$$

where the first term represents the kinetic energy and the second term the potential energy of the system. The assumed form of the potential energy means that the particles interact only pairwise through the central potential $\phi(r)$, which is a reasonable model of reality.

Various restrictions on $\phi(r)$ will be imposed later to ensure the existence of the thermodynamic limit. Intuitively, it is clear that $\phi(r)$ should have the form shown in Fig. 3.1. That is, it should be sufficiently repulsive at the origin to prevent the system from collapsing and vanish sufficiently rapidly at infinity to prevent the system from flying apart. Precise conditions are given in eqns (3.7.13) and (3.7.14).

For Hamiltonians of the form given in eqn (3.1.2), the partition function becomes

$$\begin{aligned} Z(V, N, T) &= \frac{1}{N!} \int_{-\infty}^{\infty} \cdots \int \exp\left(-\beta \sum_{i=1}^{N} \mathbf{p}_i^2/2m\right) \mathrm{d}\mathbf{p}_1 \cdots \mathrm{d}\mathbf{p}_N \\ &\quad \times \int_V \cdots \int \exp\left(-\beta \sum_{1 \leqslant i < j \leqslant N} \phi(|\mathbf{r}_i - \mathbf{r}_j|)\right) \mathrm{d}\mathbf{r}_1 \cdots \mathrm{d}\mathbf{r}_N \\ &= \frac{\lambda^N}{N!} \int_V \cdots \int \exp\left(-\beta \sum_{1 \leqslant i < j \leqslant N} \phi(|\mathbf{r}_i - \mathbf{r}_j|)\right) \mathrm{d}\mathbf{r}_1 \cdots \mathrm{d}\mathbf{r}_N \end{aligned} \tag{3.1.3}$$

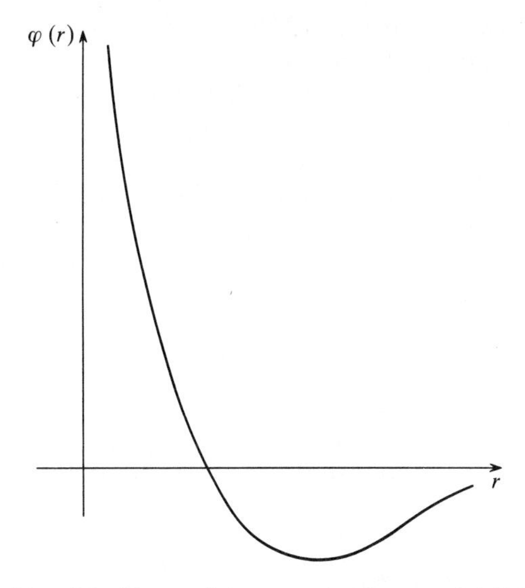

FIG. 3.1 Form of the central pair potential (3.1.2).

where

$$\lambda = (2\pi mkT)^{3/2}. \tag{3.1.4}$$

For simplicity, we will henceforth ignore the trivial multiplicative factor λ^N which arises from the kinetic energy of the system.

The above formulae are appropriate for fluid systems. To model magnetic systems we take the microscopic objects to be 'spins' which may be thought of as scalar or vector quantities $\mathbf{S}_i$ occupying fixed points in space and satisfying some constraint, such as having unit length. The phase space then becomes the set of allowed configurations of the spins, such as

$$\{S\} = \{\mathbf{S}_i | i = 1, 2, \ldots, N. \ \|\mathbf{S}_i\| = 1\}. \tag{3.1.5}$$

If $\mathscr{H}\{S\}$ is the Hamiltonian, or interaction energy, of the system in configuration $\{S\}$ the canonical partition function is defined by

$$Z(H, N, T) = C_N \int \exp(-\beta \mathscr{H}\{S\}) \mathrm{d}S \tag{3.1.6}$$

where H denotes an external magnetic field, C_N is some constant depending only on N, and the integration is taken over the configuration space of spins. For example, in the case of eqn (3.1.5) the integral in eqn (3.1.6) involves integrating each spin over the surface of the unit sphere.

In the special case where the spins in eqn (3.1.5) are scalar quantities μ_i which take the values plus or minus unity, the integral in eqn (3.1.6) becomes a sum over the 2^N allowed configurations

$$\{\mu\} = \{\mu_i = \pm 1 | i = 1, 2, \ldots, N\}. \tag{3.1.7}$$

Such a system is called an *Ising model.* We will have a lot more to say about such models in subsequent sections and chapters.

It is clear that the evaluation of the partition function (3.1.3) or (3.1.6) is, in general, a hopelessly difficult task. In the following two sections we give examples of simple one-dimensional systems where $Z(V, N, T)$ eqn (3.1.3) can be evaluated exactly. In Section 3.4, we discuss some general features of lattice models and in Section 3.5, we give a simple example where the partition function for a magnetic system (3.1.6) can be evaluated exactly. More complicated model systems which can be treated exactly are given in Chapter 6.

3.2 The Tonks gas

One of the simplest fluid systems imaginable is a gas of non-interacting hard spheres (i.e. billiard balls). This system has a Hamiltonian given by eqn (3.1.2) with a hard-core potential

$$\phi(r)=\begin{cases}\infty & \text{if } 0\leqslant r\leqslant a\\ 0 & \text{if } \quad r>a\end{cases}. \tag{3.2.1}$$

The 'infinity' in eqn (3.2.1) means that the particles cannot get closer than a distance a to each other so that the particles can be considered as hard spheres of diameter a.

For this potential the integrand in eqn (3.1.3) can be written as

$$\exp\left(-\beta\sum_{1\leqslant i<j\leqslant N}\phi(|\mathbf{r}_i-\mathbf{r}_j|)\right)=\prod_{1\leqslant i<j\leqslant N}S(|\mathbf{r}_i-\mathbf{r}_j|) \tag{3.2.2}$$

where

$$S(r)=\begin{cases}0 & \text{if } 0\leqslant r\leqslant a\\ 1 & \text{if } \quad r>a\end{cases} \tag{3.2.3}$$

The temperature, therefore, does not enter as a parameter in this model and the partition function eqn (3.1.3) is essentially the 'excluded volume' of N hard spheres in a region with volume V.

In one dimension the hard-core potential, eqn (3.2.1), corresponds to a gas of hard rods of length a. This problem was solved many years ago by Tonks (1936). The model in higher dimensions is unsolved except in the limiting case of infinite spatial dimension (Frisch *et al.* 1985).

The partition function for N hard rods on the interval $0\leqslant x\leqslant L$ is given by

$$Z(L,N,T)=\frac{1}{N!}\int_0^L\!\!\cdots\int\exp\left(-\beta\sum_{1\leqslant i<j\leqslant N}\phi(|x_j-x_i|)\right)\mathrm{d}x_1\cdots\mathrm{d}x_N \tag{3.2.4}$$

where the potential $\phi(r)$ is specified by eqn (3.2.1). The simplifying feature of one dimension is that the particles can be ordered on the line. That is, for arbitrary potentials ϕ, the integral in eqn (3.2.4) is $N!$ times the integral over the region

$$R: 0<x_1<x_2<\cdots<x_N<L. \tag{3.2.5}$$

This result is easily proved by induction using the fact that the integrand is a symmetric function of the variables $x_1, x_2, \ldots, x_N$. It is the fact that we can order particles on a line that makes exact solutions possible in one dimension. Unfortunately, there is no analogue of this manipulation in higher dimensions.

Once we have ordered the particles it is clear that when $x_{i+1}-x_i>a$, $x_j-x_i>a$ for all $j=i+1, i+2, \ldots, N$. It then follows from the representation (3.2.2) of the integrand and eqns (3.2.3) and (3.2.4) that

$$\begin{aligned} Z(L, N, T) &= \int\cdots\int_R \prod_{i=1}^{N-1} S(x_{i+1}-x_i)\mathrm{d}x_1\cdots\mathrm{d}x_N \\ &= \int\cdots\int_{R'} \mathrm{d}x_1\mathrm{d}x_2\cdots\mathrm{d}x_N \end{aligned} \tag{3.2.6}$$

where R' denotes the region

$$R'=\{(i-1)a<x_i<x_{i+1}-a|i=1, 2, \ldots, N\} \tag{3.2.7}$$

and, by definition, $x_{N+1}\equiv L+a$. Changing variables to

$$y_i=x_i-(i-1)a \qquad i=1, 2, \ldots, N \tag{3.2.8}$$

gives

$$\begin{aligned} Z(L, N, T) &= \int_0^l \mathrm{d}y_N \int_0^{y_N} \mathrm{d}y_{N-1}\cdots\int_0^{y_3}\mathrm{d}y_2\int_0^{y_2}\mathrm{d}y_1 \\ &= l^N/N! \end{aligned} \tag{3.2.9}$$

where

$$l=L-(N-1)a. \tag{3.2.10}$$

Using Stirling's formula $N!\sim N^N\mathrm{e}^{-N}$ as $N\to\infty$, we obtain in the thermodynamic limit for the free energy per particle ψ

$$\begin{aligned} -\beta\psi(v, T) &= \lim_{\substack{L, N\to\infty \\ v=L/N \text{ fixed}}} N^{-1}\ln Z(L, N, T) \\ &= 1+\ln(v-a). \end{aligned} \tag{3.2.11}$$

It is to be noted that $\psi(v, T)$ is only defined for $v>a$. When $v\leqslant a$, $Na\geqslant L$ so that at least two rods on the interval $0\leqslant x\leqslant L$ must overlap. It follows from eqn (3.2.4) that $Z(L, N, T)=0$ when $v\leqslant a$. The condition $v=a$, of course, corresponds to 'close packing'.

Various thermodynamic quantities can be derived from eqn (3.2.11). For example, the pressure and the isothermal compressibility are given, respectively, by

$$p = -\left(\frac{\partial \psi}{\partial T}\right)_v = \frac{kT}{v-a} \tag{3.2.12}$$

and

$$K_T = -\left(v\frac{\partial p}{\partial v}\right)^{-1} = (v-a)^2/vkT. \tag{3.2.13}$$

Notice that this model has the expected thermodynamic stability properties. That is, for $v > a$, $\psi(v, T)$ is a convex function of v, the pressure is a decreasing function of v, and K_T is non-negative. In other respects, however, the model is rather uninteresting. Apart from a branch-point singularity at the close packing value $v = a$, the free energy is an analytic function of v (and T). In higher dimensions it was conjectured some time ago by Kirkwood that the hard-sphere gas free energy has a singularity at a value of v greater than the close-packing specific volume corresponding to a phase transition. This is still an unsolved problem but note that, if the Kirkwood transition exists, it would correspond to a gas–solid transition rather than the usual gas–liquid condensation transition which arises from a temperature singularity.

Baxter (1980) showed that a two-dimensional lattice gas of hard hexagons has a phase transition so it seems extremely likely that the Kirkwood transition exists. Various numerical experiments (Alder and Wainwright 1962; see also Barker and Henderson 1976) also support this view.

3.3 The Takahashi gas

A simple generalization of the Tonks gas, due to Takahashi (1942), consists of a one-dimensional gas of hard rods (of length a) with interactions only between nearest-neighbour rods. Thus, the canonical partition function of the Takahashi gas is given by eqn (3.2.4) where the potential $\phi(x)$ satisfies the conditions

$$\phi(x) = \begin{cases} \infty & \text{if } \ 0 \leqslant x \leqslant a \\ 0 & \text{if } \ x > 2a \end{cases} \tag{3.3.1}$$

and is, in principle, arbitrary (but bounded) in the range $a \leqslant x \leqslant 2a$.

In order to evaluate the partition function for this model we repeat the steps leading to eqn (3.2.6) with the result that

$$Z(L, N, T) = \int_R \cdots \int \exp\left(-\beta \sum_{1 \leqslant i < j \leqslant N} \phi(x_j - x_i)\right) \mathrm{d}x_1 \mathrm{d}x_2 \cdots \mathrm{d}x_N$$

$$= \int_{R'} \cdots \int \exp\left[-\beta \sum_{i=1}^{N-1} \phi(x_{i+1} - x_i)\right] \mathrm{d}x_1 \mathrm{d}x_2 \cdots \mathrm{d}x_N \tag{3.3.2}$$

where the regions R and R' are specified in eqns (3.2.5) and (3.2.7) and in obtaining eqn (3.3.2) we have used the fact that from eqns (3.3.1) and (3.2.7),

$\phi(x_j - x_i)$ vanishes in R' unless $j=i+1$. In terms of the variables y_i in eqn (3.2.8), $x_{i+1}-x_i=y_{i+1}-y_i+a$, and hence, on comparing eqns (3.3.2) and (3.2.9), we have

$$Z(L,N,T)=\int_0^l\int_0^{y_N}\cdots\int_0^{y_2}\exp\left[-\beta\sum_{i=1}^{N-1}\phi(y_{i+1}-y_i+a)\right]\mathrm{d}y_1\cdots\mathrm{d}y_{N-1}\mathrm{d}y_N \tag{3.3.3}$$

where l is given by eqn (3.2.10).

If we now define the convolution $f*g$ of two functions f and g by

$$f*g(x)=\int_0^x f(x-y)g(y)\,\mathrm{d}y \tag{3.3.4}$$

it is easy to verify, from eqn (3.3.3), that

$$Z(L,N,T)=1*f*f*\cdots*f*1(l)\equiv Z(l) \tag{3.3.5}$$

where there are $N-1$ factors of f appearing in the convolution and f is defined by

$$f(x)=\exp[-\beta\phi(x+a)]. \tag{3.3.6}$$

The Laplace convolution formula (see Problem 1) states that, if

$$\mathscr{L}(h)=\int_0^\infty \mathrm{e}^{-sx}\,h(x)\mathrm{d}x \tag{3.3.7}$$

is the Laplace transform of a function h, then

$$\mathscr{L}(f*g)=\mathscr{L}(f)\,\mathscr{L}(g). \tag{3.3.8}$$

Using this result and considering $Z(L,N,T)$ in eqn (3.3.3) as a function $Z(l)$ of the single variable $l=L-(N-1)a$, which we now take as an integration variable, we obtain from eqns (3.3.5) and (3.3.8)

$$\int_0^\infty \mathrm{e}^{-sl}Z(l)\,\mathrm{d}l=s^{-2}[F(s)]^{N-1} \tag{3.3.9}$$

where, from eqn (3.3.6),

$$F(s)=\int_0^\infty \exp[-sx-\beta\phi(x+a)]\,\mathrm{d}x \tag{3.3.10}$$

and use has been made of the fact that $\mathscr{L}(1)=s^{-1}$.

The partition function can now be obtained by inverting the Laplace transform, giving

$$Z(L,N,T)=\frac{1}{2\pi\mathrm{i}}\int_{c-\mathrm{i}\infty}^{c+\mathrm{i}\infty}\mathrm{e}^{ls}s^{-2}[F(s)]^{N-1}\,\mathrm{d}s \tag{3.3.11}$$

where c is chosen so that all singularities of the integrand in the complex s-plane are to the left of the line $s=c$.

In the case of non-interacting hard rods (i.e. the Tonks gas) $\phi(x+a)=0$ for $x\geqslant 0$ so from eqn (3.3.10) $F(s)=s^{-1}$. The previous result, eqn (3.2.9), then follows immediately from eqn (3.3.11) by closing the contour to the left around the origin and using Cauchy's integral theorem.

In general, $Z(L, N, T)$ in eqn (3.3.11) can be evaluated asymptotically for large N (and L) with $v=L/N$ fixed, by writing

$$l=L-(N-1)a=N(v-a)+a \tag{3.3.12}$$

and evaluating the integral (3.3.11),

$$Z(L,N,T)=\frac{1}{2\pi \mathrm{i}}\int_{c-\mathrm{i}\infty}^{c+\mathrm{i}\infty}\exp\{N[(v-a)s+\ln F(s)]\}\,\mathrm{e}^{as}s^{-2}[F(s)]^{-1}\,\mathrm{d}s, \tag{3.3.13}$$

by the method of steepest descents. We thus obtain in the thermodynamic limit the expression

$$-\beta\psi(v,T)=\lim_{\substack{L,\,N\to\infty\\ v=L/N\ \text{fixed}}} N^{-1}\ln Z(L,N,T)$$

$$=(v-a)s_0+\ln F(s_0) \tag{3.3.14}$$

for the free energy $\psi(v, T)$ where the saddle point $s_0=s_0(v, T)$ is determined from

$$F'(s_0)/F(s_0)=a-v \tag{3.3.15}$$

and $F(s)$ is defined by eqn (3.3.10).

If we differentiate eqn (3.3.14) with respect to v we obtain

$$\beta p=\frac{\partial}{\partial v}(-\beta\psi)=s_0 \tag{3.3.16}$$

so that, on substituting back into eqn (3.3.15), we obtain the equation of state

$$v=a-F'(\beta p)/F(\beta p). \tag{3.3.17}$$

An alternative derivation of this result based on a direct application of Laplace's method to eqn (3.3.9) (see Problem 2) forms the basis of Problem 3. We note also from Problem 4 that p is a strictly decreasing function of v as for the Tonks gas.

In studying special cases of the Takahashi gas it is extremely important not to forget the condition on the potential stated in eqn (3.3.1), that is it vanishes when $x>2a$. It follows from eqn (3.3.10) that

$$F(s)=\int_0^a \exp(-sx-\beta\phi(x+a))\,\mathrm{d}x+s^{-1}\mathrm{e}^{-as} \tag{3.3.18}$$

and hence from eqn (3.3.17) that, so long as ϕ is bounded, v is an analytic function of p and vice versa. In other words, the Takahashi gas does not undergo a phase transition.

A common error, which seems to reoccur in the literature, stems from beginning with eqn (3.3.2) for the partition function and then ignoring the condition $\phi(x)=0$ for $x>2a$ which is essential in the reduction of the partition function to the form given in eqn (3.3.2). Further discussion on this point and other generalizations of the Tonks and Takahashi gases can be found in Lieb and Mattis (1966).

3.4 Lattice models of magnets, gases, and alloys

A typical model of a magnet consists of a set $\{S\}$ of 'spins' occupying the vertices of a regular lattice with some interaction energy or Hamiltonian $\mathscr{H}\{S\}$ depending on the configuration $\{S\}$. The canonical partition function is then defined by

$$Z=\int \exp(-\beta\mathscr{H}\{S\})\,\mathrm{d}S \tag{3.4.1}$$

where the 'integral' in eqn (3.4.1) is taken over the allowed configuration space of spins. In the special case where the configuration space is discrete or finite the integral in eqn (3.4.1) is replaced by a sum over the allowed spin configurations.

The simplest possible model of a magnet or spin system consists of a set of N scalar spins $\mu_1, \mu_2, \ldots, \mu_N$ which can take only two values, $\mu_i=+1$ corresponding to 'spin up' and $\mu_i=-1$ corresponding to 'spin down'. Such a system is usually called an 'Ising model' and the μ_i are called 'Ising spins' after E. Ising who first studied such systems (Ising 1925). We will have more to say about Ising and his model later.

Since $\mu_i=\pm 1$, $i=1, 2, \ldots, N$, the total number of configurations

$$\{\mu\}=\{\mu_1, \mu_2, \ldots, \mu_N\} \tag{3.4.2}$$

is 2^N and, if $\mathscr{H}\{\mu\}$ is the Hamiltonian or interaction energy of the system in configuration $\{\mu\}$, the canonical partition function is defined by

$$Z=\sum_{\{\mu\}} \exp(-\beta\mathscr{H}\{\mu\}) \tag{3.4.3}$$

where the sum is taken over all 2^N configurations $\{\mu\}$. To be more specific, assuming that we have only pairwise interactions between the spins and an interaction between each spin and an external magnetic field H, the Hamiltonian may be written as

$$\mathscr{H}\{\mu\}=-\sum_{1\leq i<j\leq N} J_{ij}\mu_i\mu_j - H\sum_{i=1}^{N}\mu_i \tag{3.4.4}$$

where J_{ij} denotes a coupling constant between spins labelled i and j such that the interaction energy between these spins is $-J_{ij}$ if the spins are parallel ($\mu_i\mu_j = +1$) and $+J_{ij}$ if the spins are antiparallel ($\mu_i\mu_j = -1$). The sign convention here means that a positive coupling constant favours, by virtue of having lower energy, a 'ferromagnetic' or parallel alignment of spins whereas a negative coupling constant favours an 'antiferromagnetic' or antiparallel alignment of spins. The partition function (3.4.3) is usually considered as a function of H, N, and T and, in certain circumstances, as a function of the set of coupling constants $\{J_{ij}\}$ and its associated symmetries, and of course the dimension of the underlying lattice.

A natural generalization of the Ising model consists of replacing the Ising spins $\mu_i = \pm 1$ by n-dimensional unit vectors $\mathbf{S}_i$. The configuration space is then

$$\{S\} = \{\mathbf{S}_i \,|\, i = 1, 2, \ldots, N.\ \|\mathbf{S}_i\| = 1\} \tag{3.4.5}$$

and, for a system with Hamiltonian $\mathscr{H}\{S\}$, the canonical partition function is given by

$$Z = \underset{\{\|\mathbf{S}_i\| = 1\}}{\int \cdots \int} \exp(-\beta\mathscr{H}\{S\})\, \mathrm{d}\mathbf{S}_1 \cdots \mathrm{d}\mathbf{S}_N \tag{3.4.6}$$

where, with obvious notation, the configuration integral is taken over the N unit spheres $\|\mathbf{S}_i\| = 1$, $i = 1, 2, \ldots, N$.

If we have only pairwise interactions and an external magnetic field $\mathbf{H}$, the natural generalization of the Ising model Hamiltonian (3.4.4) for the n-vector model is

$$\mathscr{H}\{S\} = -\sum_{1 \leqslant i < j \leqslant N} J_{ij}\mathbf{S}_i \cdot \mathbf{S}_j - \mathbf{H} \cdot \sum_{i=1}^{N} \mathbf{S}_i \tag{3.4.7}$$

where

$$\mathbf{A} \cdot \mathbf{B} = \sum_{\alpha=1}^{n} A_\alpha B_\alpha \tag{3.4.8}$$

denotes the scalar product of two n-dimensional vectors $\mathbf{A} = (A_1, A_2, \ldots, A_n)$ and $\mathbf{B} = (B_1, B_2, \ldots, B_n)$.

It is to be noted that, since $\|\mathbf{S}_i\| = 1$, the n-vector model reduces to the Ising model when $n = 1$. When $n = 2$, the model is called the planar classical Heisenberg or plane rotor model and, when $n = 3$, we have the usual classical Heisenberg model. We shall see later that the model has some interesting properties in the limit $n \to \infty$.

The binary nature of the Ising spin variables permits other possible interpretations of the Ising model. We could, for example, think of the model as a 'lattice gas' with 'spin up' corresponding to an occupied site and 'spin down' corresponding to an unoccupied site, or as a model of a binary alloy with spin up and spin down corresponding to a lattice site being occupied by an A or a B atom, respectively.

The term lattice gas was first introduced by Lee and Yang (1952) although the interpretation of the Ising model as a gas was, in fact, known earlier.

The definition of the lattice gas and its precise connection with the Ising model is as follows.

Consider a lattice of V sites (V being thought of as a 'volume') and a collection of N particles that can occupy these sites with the restriction that not more than one particle can occupy a given site. To each site we assign a variable t_i defined by

$$t_i = \begin{cases} 1 & \text{if site } i \text{ is occupied} \\ 0 & \text{if site } i \text{ is unoccupied.} \end{cases} \tag{3.4.9}$$

Since the total number of particles is N, we have the constraint

$$\sum_{i=1}^{V} t_i = N. \tag{3.4.10}$$

We assume that the interaction energy between sites i and j is $-\phi_{ij}$ if sites i and j are occupied and zero otherwise. It follows that the total interaction energy $E\{t\}$ in a given configuration

$$\{t\} = \{t_1, t_2, \ldots, t_V\} \tag{3.4.11}$$

is given by

$$E\{t\} = -\sum_{1 \leqslant i < j \leqslant V} \phi_{ij} t_i t_j. \tag{3.4.12}$$

The canonical partition function is then given by

$$Z(V, N, T) = \sum_{\{t\}}{}' \exp(-\beta E\{t\}) \tag{3.4.13}$$

where the primed sum denotes the sum over all configurations satisfying the constraint (3.4.10). To remove the restriction on the sum in eqn (3.4.13) it is convenient to consider the grand-canonical partition function

$$\begin{aligned} Z_G(V, T, z) &= \sum_{N=0}^{\infty} z^N Z(V, N, T) \\ &= \sum_{\{t\}} \exp\left(\ln z \sum_{i=1}^{V} t_i + \beta \sum_{1 \leqslant i < j \leqslant V} \phi_{ij} t_i t_j \right) \end{aligned} \tag{3.4.14}$$

where use has been made of eqns (3.4.10), (3.4.12), and (3.4.13), and the sum over $\{t\}$ in eqn (3.4.14) is now over all configurations.

To set up a precise relationship between the lattice gas and the Ising magnet we define the spin variable μ_i by

$$t_i = \tfrac{1}{2}(1 - \mu_i) \tag{3.4.15}$$

which takes the values $\mu_i = +1$ or -1 when $t_i = 0$ or 1, respectively. Substituting eqn (3.4.15) into eqn (3.4.14) shows after an elementary calculation that

$$Z_G(V,T,z)=\exp\left(\frac{V}{2}\ln z+\beta\sum_{1\leqslant i<j\leqslant V}J_{ij}\right)$$

$$\times\sum_{\{\mu\}}\exp\left(\beta\sum_{1\leqslant i<j\leqslant V}J_{ij}\mu_i\mu_j+\beta\sum_{i=1}^{V}h_i\mu_i\right) \tag{3.4.16}$$

where

$$J_{ij}=\phi_{ij}/4 \tag{3.4.17}$$

and

$$\beta h_i=-\frac{1}{2}\ln z-\beta\left(\sum_{j=i+1}^{V}J_{ij}+\sum_{j=1}^{i-1}J_{ji}\right) \tag{3.4.18}$$

where the first and second sums in eqn (3.4.18) are replaced by zero when $i=V$ and $i=1$, respectively.

Apart from the trivial multiplicative factor on the right-hand side of eqn (3.4.16), $Z_G(V, T, z)$ is precisely the canonical partition function for an Ising model having V sites with coupling constants J_{ij} between spins at sites i and j and with an external field h_i applied to the spin on site i.

It is now straightforward to set up relationships between thermodynamic quantities for a lattice gas and the corresponding Ising magnetic model. For example, the *density* $\rho=\langle N\rangle/V$ of the lattice gas, defined by

$$\rho=z\frac{\partial}{\partial z}(V^{-1}\ln Z_G(V,T,z))$$

$$=[Z_G(V,T,z)]^{-1}\sum_{\{t\}}\left(V^{-1}\sum_{i=1}^{V}t_i\right)\exp\left(\ln z\sum_{i=1}^{V}t_i+\beta\sum_{1\leqslant i<j\leqslant V}\phi_{ij}t_it_j\right), \tag{3.4.19}$$

is obviously related to the magnetization per spin of the corresponding Ising model

$$m=\left\langle V^{-1}\sum_{i=1}^{V}\mu_i\right\rangle \tag{3.4.20}$$

through

$$\rho=\tfrac{1}{2}(1-m) \tag{3.4.21}$$

which is easily obtained by noting eqn (3.4.15). Other relations are given in Problem 6.

We conclude this section with a brief mention of the binary-alloy interpretation of the Ising model.

If we have an alloy composed of A and B atoms we can reinterpret the lattice gas variables as

$$t_i=\begin{cases}1 & \text{if site } i \text{ is occupied by an A atom}\\ 0 & \text{if site } i \text{ is occupied by a B atom.}\end{cases} \tag{3.4.22}$$

Then if $-\phi_{ij}^{AA}$, $-\phi_{ij}^{BB}$, and $-\phi_{ij}^{AB}$ denote the interaction energies between sites i and j when they are occupied, respectively, by two A atoms, two B atoms, and an A and a B atom, the total interaction energy $E\{t\}$ in a given configuration $\{t\}$ of A and B atoms is given by

$$E\{t\} = -\sum_{1 \leq i < j \leq V} \{\phi_{ij}^{AA} t_i t_j + \phi_{ij}^{BB}(1-t_i)(1-t_j) + \phi_{ij}^{AB}[t_i(1-t_j)+(1-t_i)t_j]\}. \quad (3.4.23)$$

It is left as an exercise to set up a precise correspondence between the binary-alloy model and the Ising model. It is almost obvious, for example, that the binary alloy with equal numbers of A and B atoms corresponds to an Ising model with zero external magnetic field.

3.5 The one-dimensional *n*-vector model

For simplicity, let us first consider the one-dimensional Ising model ($n=1$) with nearest-neighbour interactions only and in zero external magnetic field. For such a chain of N Ising spins, the interaction energy $E\{\mu\}$ in a given configuration $\{\mu\}$ is

$$E\{\mu\} = -\sum_{i=1}^{N-1} J_i \mu_i \mu_{i+1} \quad (3.5.1)$$

and the canonical partition function, eqn (3.4.3), becomes

$$Z(N,T) = \sum_{\{\mu\}} \exp\left[\beta \sum_{i=1}^{N-1} J_i \mu_i \mu_{i+1}\right]$$

$$= \sum_{\mu_1 = \pm 1} \cdots \sum_{\mu_N = \pm 1} \prod_{i=1}^{N-1} \exp(\beta J_i \mu_i \mu_{i+1}). \quad (3.5.2)$$

If we sum over $\mu_N = \pm 1$ and use the fact that

$$\sum_{\mu_N = \pm 1} \exp(\beta J_{N-1} \mu_{N-1} \mu_N) = \exp(\beta J_{N-1} \mu_{N-1}) + \exp(-\beta J_{N-1} \mu_{N-1})$$

$$= 2 \cosh(\beta J_{N-1} \mu_{N-1})$$

$$= 2 \cosh(\beta J_{N-1}), \quad (3.5.3)$$

where, in the last step, we have used the fact that $\mu_{N-1} = \pm 1$ and that $\cosh x$ is an even function of x, it follows immediately from eqns (3.5.2) and (3.5.3) that

$$Z(N,T) = 2 \cosh(\beta J_{N-1}) Z(N-1,T). \quad (3.5.4)$$

By summing successively over $\mu_{N-1}, \mu_{N-2}, \ldots, \mu_2, \mu_1$ or, equivalently, iterating the recurrence relation (3.5.4) and using the fact that

$$Z(2,T) = \sum_{\mu_1=\pm 1}\sum_{\mu_2=\pm 1} \exp(\beta J_1 \mu_1 \mu_2)$$
$$= \sum_{\mu_1=\pm 1} 2\cosh(\beta J_1)$$
$$= 4\cosh(\beta J_1), \tag{3.5.5}$$

we arrive at the exact form of the partition function,

$$Z(N,T) = 2\prod_{i=1}^{N-1} 2\cosh(\beta J_i). \tag{3.5.6}$$

In the special case where all the coupling constants J_i are equal, to J say, the free energy per spin in the thermodynamic limit ψ is given by

$$-\beta\psi = \lim_{N\to\infty} N^{-1}\ln Z(N,T)$$
$$= \ln(2\cosh\beta J). \tag{3.5.7}$$

Notice that ψ is an analytic function of β so again there is no phase transition.

The above derivation relies heavily on eqn (3.5.3) and the fact that the right-hand side of this equation is independent of μ_{N-1}. In the presence of an external magnetic field, the μ_{N-1}-dependence remains when we sum over μ_N in eqn (3.5.2) so the above method does not work in this case. Another method which works in the presence of a field, and which enables us to compute derived quantities such as the magnetization, is given in Chapter 6. There is however another quantity, the pair-correlation function in zero field, which gives us a measure of the degree of magnetic ordering, and which can be computed by a slight variation of the above method.

The pair-correlation function in general is defined by

$$\langle \mu_k \mu_l \rangle = \sum_{\{\mu\}} \mu_k \mu_l \exp(-\beta E\{\mu\}) \Big/ \sum_{\{\mu\}} \exp(-\beta E\{\mu\}). \tag{3.5.8}$$

In zero field with nearest-neighbour interactions only and interaction energy (3.5.1), it is not difficult to show (see Problem 8) that, when $J_i = J$,

$$\langle \mu_k \mu_l \rangle = [\tanh(\beta J)]^{|k-l|}. \tag{3.5.9}$$

At zero temperature ($\beta\to\infty$) the system is completely ordered ($\langle \mu_k \mu_l \rangle = 1$) but at all finite temperatures the pair-correlation function decays exponentially fast with power proportional to the distance between the two spins. In this case we say that we have no *long-range order*.

The n-vector model for $n>1$ in zero magnetic field and with nearest-neighbour interactions only can be treated in a similar way to the above.

Thus, in order to calculate the canonical partition function

$$Z(N,T)=\int_{\{\|\mathbf{S}_i\|=1\}}\cdots\int \exp\left(\beta J\sum_{i=1}^{N-1}\mathbf{S}_i\cdot\mathbf{S}_{i+1}\right)\mathrm{d}\mathbf{S}_1\cdots\mathrm{d}\mathbf{S}_N, \qquad (3.5.10)$$

we integrate successively over the spins $\mathbf{S}_N, \mathbf{S}_{N-1}, \ldots, \mathbf{S}_1$ and use the result (see Problem 9)

$$\int_{\|\mathbf{S}\|=1} \exp(\boldsymbol{\alpha}\cdot\mathbf{S})\,\mathrm{d}\mathbf{S}=2\pi^{n/2}\frac{I_{\frac{n}{2}-1}(\|\boldsymbol{\alpha}\|)}{(\|\boldsymbol{\alpha}\|/2)^{\frac{n}{2}-1}} \qquad (3.5.11)$$

where $I_\mu(x)$ is the modified Bessel function of the first kind of order μ defined by

$$I_\mu(x)=(x/2)^\mu(2\pi\mathrm{i})^{-1}\int_{c-\mathrm{i}\infty}^{c+\mathrm{i}\infty}\exp(z+x^2/4z)z^{-(\mu+1)}\,\mathrm{d}z \qquad (c>0) \qquad (3.5.12)$$

and $\boldsymbol{\alpha}$ is an arbitrary n-dimensional vector. In the particular case where $\boldsymbol{\alpha}=\beta J\mathbf{S}'$ and $\mathbf{S}'$ is a unit (spin) vector, the right-hand side of eqn (3.5.11) is independent of $\mathbf{S}'$. It then follows straightforwardly that the canonical partition function (3.5.10) is given by

$$Z(N,T)=\frac{2\pi^{n/2}}{\Gamma(n/2)}\left[2\pi^{n/2}I_{\frac{n}{2}-1}(\beta J)/(\beta J/2)^{\frac{n}{2}-1}\right]^{N-1} \qquad (3.5.13)$$

from which one can then obtain the free energy and derived field-independent thermodynamic quantities.

The pair-correlation function, for example, is given by (see Problem 10)

$$\langle\mathbf{S}_k\cdot\mathbf{S}_l\rangle=[I_{\frac{n}{2}}(\beta J)/I_{\frac{n}{2}-1}(\beta J)]^{|k-l|} \qquad (3.5.14)$$

and, again, there is zero long-range order at all finite temperatures and no phase transition.

3.6 The thermodynamic limit for a lattice-model system

As we saw in Chapter 2 the various distributions in equilibrium statistical mechanics only give equivalent thermodynamic descriptions in the thermodynamic or bulk limit of an infinite system. And as we will see in Chapter 4 we need to take the thermodynamic limit as a prelude to a precise and rigorous discussion of phase transitions.

In this and the following section we consider the existence of the thermodynamic limit for a simple spin model and for continuum systems with simple potentials. The proofs presented here are designed to illustrate general techniques so the models we consider are somewhat specialized. A more thorough and exhaustive discussion of the existence of the thermodynamic limit can be found in Ruelle (1969).

For simplicity, we consider first a one-dimensional Ising model consisting of $N=2^n$ spins and with interaction energy

$$E_n\{\mu\} = -\sum_{1\leqslant i<j\leqslant N} J(j-i)\,\mu_i\mu_j. \tag{3.6.1}$$

With this form of the coupling constants the model is translationally invariant. Restrictions on $J(k)$ will be imposed later.

The partition function is defined by

$$\begin{aligned} Z_n &= Z(2^n, T) \\ &= \sum_{\{\mu\}} \exp(-\beta E_n\{\mu\}) \end{aligned} \tag{3.6.2}$$

and the free energy is readily computed from the quantities f_n defined by

$$f_n = 2^{-n} \ln Z_n. \tag{3.6.3}$$

Our aim is to establish conditions on $J(k)$ to ensure the existence of the limiting free energy per spin ψ given by

$$-\beta\psi = \lim_{n\to\infty} f_n. \tag{3.6.4}$$

We will show that the sequence $\{f_n\}$ is always monotonic. Conditions on $J(k)$ which force the sequence to be bounded will then ensure its convergence and the required result (3.6.4).

In order to show that the sequence $\{f_n\}$ is monotonic we first divide the chain of 2^n spins in two and write the interaction energy, (3.6.1) in the form

$$E_n\{\mu\} = E_{n-1}\{\sigma\} + E_{n-1}\{\sigma'\} + E_{\text{I}}\{\sigma, \sigma'\} \tag{3.6.5}$$

where $\{\sigma\}$ and $\{\sigma'\}$ denote the two halves of the configuration $\{\mu\}$, i.e.

$$\left.\begin{aligned} \sigma_i &= \mu_i \\ \sigma'_i &= \mu_{2^{n-1}+i} \end{aligned}\right\} \quad i=1,2,\ldots,2^{n-1} \tag{3.6.6}$$

and we have used translation invariance to write the interaction energies of the two halves of the chain as $E_{n-1}\{\sigma\}$ and $E_{n-1}\{\sigma'\}$ where $E_n\{\mu\}$ is defined (for any n and $\{\mu\}$) by eqn (3.6.1). The contribution $E_{\text{I}}\{\mu\}$ to $E_n\{\mu\}$ in eqn (3.6.5) represents the interaction energy between the two halves of the original chain, i.e.

$$\begin{aligned} E_{\text{I}}\{\sigma, \sigma'\} &= -\sum_{i=1}^{2^{n-1}} \sum_{j=2^{n-1}+1}^{2^n} J(j-i)\,\mu_i\mu_j \\ &= -\sum_{i=1}^{2^{n-1}} \sum_{j=1}^{2^{n-1}} J(j-i+2^{n-1})\,\sigma_i\sigma'_j. \end{aligned} \tag{3.6.7}$$

We now define the normalized probability distribution $P\{\sigma,\sigma'\}$ by

$$P\{\sigma,\sigma'\}=Z_{n-1}^{-2}\exp(-\beta E_{n-1}\{\sigma\}-\beta E_{n-1}\{\sigma'\}) \tag{3.6.8}$$

where Z_n is the partition function for 2^n spins defined by eqn (3.6.2). With the above decomposition of $E\{\mu\}$ in eqn (3.6.5) and the notation in eqn (3.6.8), Z_n in eqn (3.6.2) can be expressed as

$$\begin{aligned}Z_n&=Z_{n-1}^2\sum_{\{\sigma,\sigma'\}}P\{\sigma,\sigma'\}\exp(-\beta E_{\text{I}}\{\sigma,\sigma'\})\\&=Z_{n-1}^2\langle\exp(-\beta E_{\text{I}})\rangle\end{aligned} \tag{3.6.9}$$

where the angular brackets denote an expectation value with respect to the distribution P.

Written in the form of eqn (3.6.9) we can now use Jensen's inequality (see Problem 11)

$$\langle\exp X\rangle\geqslant\exp\langle X\rangle \tag{3.6.10}$$

which holds for any expectation value $\langle\cdot\rangle$ and random variable X, to obtain

$$Z_n\geqslant Z_{n-1}^2\exp\langle-\beta E_{\text{I}}\rangle. \tag{3.6.11}$$

Using now the fact that $E_{n-1}\{\sigma\}$ and hence $P\{\sigma,\sigma'\}$ is symmetric in σ, i.e. unchanged by reversing the signs of all spins $\sigma_1,\ldots,\sigma_{2^{n-1}}$, whereas from eqn (3.6.7)

$$E_{\text{I}}\{-\sigma,\sigma'\}=-E_{\text{I}}\{\sigma,\sigma'\}, \tag{3.6.12}$$

we immediately deduce that $\langle E_{\text{I}}\rangle=0$ and hence

$$Z_n\geqslant Z_{n-1}^2. \tag{3.6.13}$$

It then follows from eqn (3.6.3) that,

$$f_n\geqslant f_{n-1}, \tag{3.6.14}$$

i.e. the sequence $\{f_n\}$ is monotonic increasing. The required limit in eqn (3.6.4) will then exist provided the sequence $\{f_n\}$ is bounded above.

From eqns (3.6.1) and (3.6.2) it is obvious, since there are 2^n spins and $\mu_i=\pm1$, that with $N=2^n$,

$$\begin{aligned}E_n\{u\}&\geqslant-\sum_{1\leqslant i<j\leqslant N}|J(j-i)|\\&\geqslant-N\sum_{k=1}^{N}|J(k)|\end{aligned} \tag{3.6.15}$$

and hence that

$$Z_n\leqslant 2^N\exp\left(N\beta\sum_{k=1}^{N}|J(k)|\right) \tag{3.6.16}$$

where the prefactor represents the total number of configurations. It then follows from eqn (3.6.3) that

$$f_n \leqslant \ln 2 + A\beta \tag{3.6.17}$$

is bounded above provided

$$A = \sum_{k=1}^{\infty} |J(k)| < \infty. \tag{3.6.18}$$

The condition given in eqn (3.6.18) is sufficient, and close to being necessary, for the existence of the thermodynamic limit. Thus, if the sum in eqn (3.6.18) is divergent, there will presumably be some configuration (e.g. all $\mu_i = +1$ when $J(k) \geqslant 0$) for which $E\{\mu\}/N$, and hence Z_N, will diverge when $N \to \infty$.

The above argument is easily extended to higher dimensions and to other spin models. In such cases it is easy to show that the thermodynamic limit exists when

$$\sum_{\mathbf{r}} |J(\mathbf{r})| < \infty \tag{3.6.19}$$

where the sum is taken over all sites of the infinite lattice on which the model in question is defined.

3.7 The thermodynamic limit for a continuum-model system

For continuum systems the thermodynamic limit is more complicated since we need to take the size of the system as well as the number of particles to infinity while holding the density fixed.

To be more precise, consider a sequence of three-dimensional domains Ω_k, with volumes V_k, $k = 1, 2, \ldots$ satisfying $V_{k+1} > V_k$, containing N_k particles with fixed density $\rho = N_k/V_k$. The canonical partition function for the kth domain is defined by

$$\begin{aligned} Z_k &= Z(N_k, \Omega_k, T) \\ &= (N_k!)^{-1} \int \cdots \int_{\Omega_k} \exp\left(-\beta \sum_{1 \leqslant i < j \leqslant N_k} \phi(|\mathbf{r}_i - \mathbf{r}_j|)\right) \mathrm{d}\mathbf{r}_1 \cdots \mathrm{d}\mathbf{r}_{N_k} \end{aligned} \tag{3.7.1}$$

and the corresponding free energy per particle ψ_k by

$$f_k = -\beta\psi_k = N_k^{-1} \ln Z_k. \tag{3.7.2}$$

The question is: For what potentials ϕ and for what domains Ω_k does the limit

$$\lim_{k \to \infty} f_k = f(\rho, T) \tag{3.7.3}$$

exist?

We will not concern ourselves here with the domain part of this question except to say that almost any shapes for the Ω_k will do provided their surface areas do not increase too rapidly (more than $V_k^{2/3}$, for example, is too fast). The potential part of the question is clearly more interesting.

For simplicity, we will consider the special case where the potential ϕ has the form shown in Fig. 3.2, i.e.

$$\phi(r) = \begin{cases} \infty & \text{if } 0 \leqslant r \leqslant a \\ <0 & \text{if } a < r < b \\ 0 & \text{if } \quad r \geqslant b \\ > -c & \text{if } \quad r > a. \end{cases} \tag{3.7.4}$$

In other words, the potential has a hard core of diameter a, and an attractive tail of finite range b. Notice that the 'hard-core' condition simply means that the domain of integration in eqn (3.7.1) is restricted to $|\mathbf{r}_i - \mathbf{r}_j| > a$ for all $i \neq j = 1, 2, \ldots, N$. Also, in the special case $b = a$, the system reduces to a gas of hard spheres of diameter a.

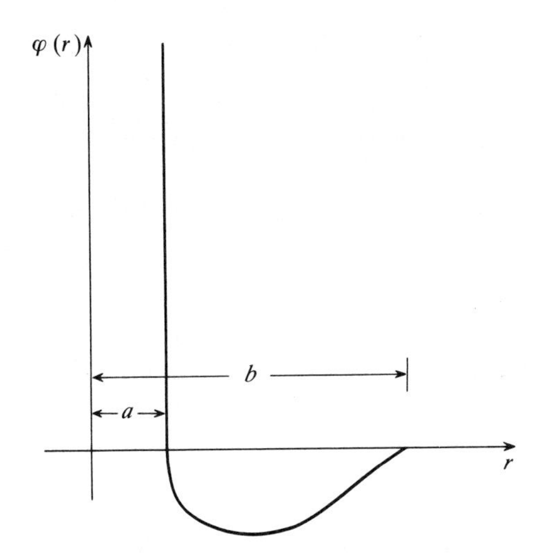

FIG. 3.2 The van Hove potential (3.7.4).

We call a potential of the form shown in Fig. 3.2 a van Hove potential, after Leon van Hove who was the first (van Hove 1949) to prove that the limit eqn (3.7.3) exists for this type of potential. Unfortunately, van Hove's proof is incomplete. The argument, however, has since been tightened up and generalized by Ruelle (1963), Fisher (1964*a*), and others. The proof given here is due to Ruelle (1963) (see also Ruelle 1969).

Before we can proceed we must specify the sequence of domains Ω_k, $k = 1, 2, \ldots$. For simplicity, we take a *standard sequence of cubes* defined as follows. Start with a cube Ω_1, with 'free volume' V_1, and walls of thickness $a/2$. Proceeding inductively, we construct the domain Ω_{k+1} by placing eight Ω_k cubes with free volumes V_k and walls of thickness $a/2$, in a larger cube with free volume $V_{k+1}=8V_k$ and walls of thickness $a/2$, as shown in Fig. 3.3. The number of particles in Ω_{k+1} is $N_{k+1}=8N_k$ and the partition function for the kth domain is defined by eqn (3.7.1).

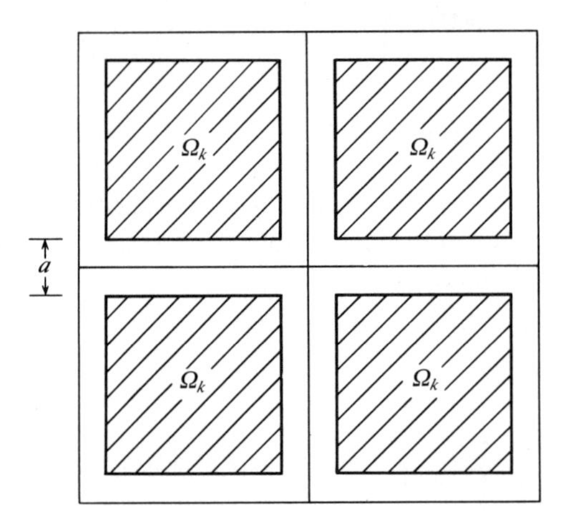

FIG. 3.3 The standard sequence of cubes $\{\Omega_k\}$.

Our strategy now, as in the previous section, is to show that the sequence of 'free energies' $\{f_k\}$ is monotonic and bounded. This is sufficient to establish that the thermodynamic limit eqn (3.7.2) exists. The artificial construction of walls or 'corridors' in our standard sequence of cubes is designed to facilitate the monotonicity argument. With some degree of care it can be eliminated by noting that the fraction of the total volume contained in corridors approaches zero in the thermodynamic limit.

In order to obtain a lower bound for Z_k defined by eqn (3.7.1) we first decrease the domain of integration by restricting N_{k-1} of the $N_k=8N_{k-1}$ particles to be in each of the eight Ω_{k-1} cubes making up Ω_k. Since the integrand is positive, this decreases the right-hand side of eqn (3.7.1). Next, we note that, when particles i and j are in different Ω_{k-1} cubes, they are separated by a distance of at least the width a of the corridors. In such a case, from eqn (3.7.4), $\phi(|\mathbf{r}_i-\mathbf{r}_j|)\leqslant 0$ and hence the contribution to the integrand in eqn (3.7.1) from such interactions is greater than or equal to unity. Elimination of interactions between particles in different Ω_{k-1} cubes thus further decreases the value of the right-hand side of eqn (3.7.1) which then decouples into a product of eight identical integrals. Noting, in the first step, that there are $(8N_{k-1})!/(N_{k-1}!)^8$ ways of arranging the N_k particles in eight identical

Ω_{k-1} cubes, it follows from eqn (3.7.1) that

$$Z_k \geqslant (Z_{k-1})^8. \tag{3.7.5}$$

From the definition of f_k given in eqn (3.7.2) we then obtain the required monotonicity result

$$f_k \geqslant f_{k-1}. \tag{3.7.6}$$

The sequence $\{f_k\}$ will then have a limit provided it is bounded above. This is the case in general if the potential $\phi(r)$ satisfies the *stability condition*

$$\sum_{1 \leqslant i < j \leqslant N} \phi(|\mathbf{r}_i - \mathbf{r}_j|) > -BN \qquad \text{for all } N \text{ and configurations} \tag{3.7.7}$$

$$\{\mathbf{r}\} = \{\mathbf{r}_1, \ldots, \mathbf{r}_N\}$$

where B is a positive constant independent of N. Thus, from eqn (3.7.7) and the definition eqn (3.7.1), we have

$$Z_k < (N_k!)^{-1} V_k^{N_k} \exp(\beta B N_k). \tag{3.7.8}$$

Now, since

$$\ln N! > N \ln N - N \tag{3.7.9}$$

and we are holding the density $\rho = N_k / V_k$ fixed, it follows from eqns (3.7.2) and (3.7.8) that

$$f_k < 1 + \beta B - \ln \rho \qquad \text{for all } k = 1, 2, \ldots \tag{3.7.10}$$

which gives the required upper bound.

In order to complete our proof of the existence of the thermodynamic limit (3.7.3) we need only to show now that the van Hove potential (3.7.4) satisfies the stability condition (3.7.7). From the condition $\phi(r) = 0$ for $r \geqslant b$ in eqn (3.7.4), it follows that only a finite number of particles, the number $s(a; b)$ of spheres of diameter a which can be packed into a sphere of diameter b, can interact with a given particle. In other words, since from eqn (3.7.4) $\phi(r) > -c$ for all r,

$$\sum_{j=1}^{N} \phi(|\mathbf{r}_i - \mathbf{r}_j|) > -cs(a; b) \qquad \text{for all } |\mathbf{r}_i - \mathbf{r}_j| > a$$

$$i, j = 1, 2, \ldots, N. \tag{3.7.11}$$

It then follows that

$$\sum_{1 \leqslant i < j \leqslant N} \phi(|\mathbf{r}_i - \mathbf{r}_j|) > -Ncs(a; b) \qquad \text{for all (allowed) configurations } \{\mathbf{r}\} \tag{3.7.12}$$

which is the required result, eqn (3.7.7).

It should be noted that we could have arrived at the monotonicity result, eqn (3.7.6), by replacing the condition $\phi(r) = 0$ for $r \geqslant b$ in eqn (3.7.4) by the so-called *weak tempering condition* $\phi(r) \leqslant 0$ for $r > R$. The above proof of monotonicity then holds if the cubes Ω_k in the standard sequence have walls of thickness $R/2$ instead of $a/2$.

As mentioned previously, the artificially constructed corridors in the standard sequence of cubes can be eliminated but at the expense of adding 'surface terms' to the inequality (3.7.6). With some degree of care one can show that these extra terms make a negligible contribution and the proof goes through essentially as above without too much difficulty. The interested reader is referred to Fisher (1964*a*) and Ruelle (1969) for details.

In order for the thermodynamic limit to exist for more general potentials we essentially need two conditions on the potential; one for large distances to give the monotonicity result (3.7.6) and one at small distances to make the potential stable. Sufficient conditions given by Fisher (1964*a*) are

$$\left.\begin{aligned} &\text{(a)}\quad \phi(r) > -c \\ &\text{(b)}\quad |\phi(r)| \leqslant A/r^{d+\varepsilon} \qquad \text{as } r \to \infty \end{aligned}\right\} \tag{3.7.13}$$

and

$$\text{(c)}\quad \phi(r) > A'/r^{d+\varepsilon'} \qquad \text{as } r \to 0 \tag{3.7.14}$$

where d is the dimensionality of the system and A, A', ε, and ε' are arbitrary positive numbers. Ruelle and Fisher have shown that under these conditions the thermodynamic limit for the grand-canonical distribution exists and gives equivalent thermodynamics to the canonical distribution in this limit. Similar results have been established by Griffiths (1965*b*) for the microcanonical distribution.

3.8 Convexity of the free energy and thermodynamic stability

In order to establish a complete connection with thermodynamics we need to show not only that the thermodynamic limit exists but also that the resulting thermodynamic description is stable. That is, as discussed in Section 1.6, quantities such as the compressibility, susceptibility, and specific heat, for example, are non-negative. These properties are a consequence of convexity of the free energy, which is easily established by a suitable modification of the arguments (and their generalizations) given in the previous two sections. Here we discuss only the fluid or continuum case. The corresponding properties for magnetic systems are posed as a sequence of problems at the end of this chapter.

In order to establish that the free energy is a convex function of the specific volume, at least for systems with a van Hove potential (eqn (3.7.4)), we proceed as before with the standard sequence of cubes $\{\Omega_k\}$ but now, in the part of the argument where we reduce the domain of integration, we restrict $N_{k-1}^{(1)}$ of the particles to four of the Ω_{k-1} cubes and $N_{k-1}^{(2)}$ of the particles to the remaining four cubes making up Ω_k. Instead of keeping only the total density fixed, we now keep $\rho_1 = N_{k-1}^{(1)}/V_{k-1}$ and $\rho_2 = N_{k-1}^{(2)}/V_{k-1}$ fixed. Eliminating interactions, as before, between different Ω_{k-1} cubes and taking account of appropriate combinatorial factors we obtain, instead of eqn (3.7.5), the inequality

$$Z_k(\Omega_k, N_k, T) \geqslant [Z_{k-1}(\Omega_{k-1}, N_{k-1}^{(1)}, T)]^4 [Z_{k-1}(\Omega_{k-1}, N_{k-1}^{(2)}, T)]^4 \quad (3.8.1)$$

where

$$N_k = 4N_{k-1}^{(1)} + 4N_{k-1}^{(2)}. \quad (3.8.2)$$

If we now define the 'free energy per unit volume' for the kth domain with density $\rho = N_k/V_k$ by

$$g_k(\rho) = V_k^{-1} \ln Z_k(\Omega_k, N_k, T), \quad (3.8.3)$$

it follows easily from eqns (3.8.1)–(3.8.3) that

$$g_k\left(\frac{\rho_1 + \rho_2}{2}\right) \geqslant \frac{1}{2}[g_{k-1}(\rho_1) + g_{k-1}(\rho_2)]. \quad (3.8.4)$$

Since g_k and f_k, defined by eqn (3.7.2), are related through

$$g_k = \frac{N_k}{V_k} f_k = \rho f_k \quad (3.8.5)$$

and $\lim f_k = f$ exists, it follows that $\lim g_k = g$ exists and, on taking limits on both sides of eqn (3.8.4), that

$$g\left(\frac{\rho_1 + \rho_2}{2}\right) \geqslant \frac{1}{2}[g(\rho_1) + g(\rho_2)]. \quad (3.8.6)$$

From this inequality we deduce that $g(\rho)$ *is a concave function of the density.*

The limiting functions f and g, from eqn (3.8.5), are related through

$$g(\rho) = v^{-1} f(v) \quad (3.8.7)$$

where we consider f to be a function of the specific volume $v = \rho^{-1}$. It is not difficult to show (see Problem 14) that f is also a concave function of v and hence, from eqn (3.7.2), that the limiting free energy per particle

$$\psi(v) = -\lim_{k \to \infty} \beta^{-1} f_k = -kTf(v) \quad (3.8.8)$$

is a convex function of v, i.e.

$$\psi\left(\frac{v_1 + v_2}{2}\right) \leqslant \frac{1}{2}[\psi(v_1) + \psi(v_2)]. \quad (3.8.9)$$

In order to proceed with questions of thermodynamic stability we need to note certain properties of convex functions. (For more detail the reader is referred to Hardy *et al.* (1964).)

1. A bounded convex function is continuous.
2. A continuous convex function ψ satisfies

$$\psi\left(\sum_{i=1}^{n} \omega_i v_i\right) \leqslant \sum_{i=1}^{n} \omega_i \psi(v_i) \tag{3.8.10}$$

for all non-negative ω_i such that $\sum_{i=1}^{n} \omega_i = 1$.

3. $d\psi/dv$ exists almost everywhere. Left- and right-hand derivatives exist everywhere and both are non-decreasing functions of v. The right-hand derivative is also not less than the left-hand derivative.

Convexity, as defined by eqn (3.8.9) has the geometrical interpretation shown in Fig. 3.4, i.e. the value of a convex function at the midpoint of an arc is not greater than the arithmetic mean of the values of the function at the two end-points of the arc. Other properties of convex functions (see Problem 3, Chapter 1) can be deduced from Fig. 3.4.

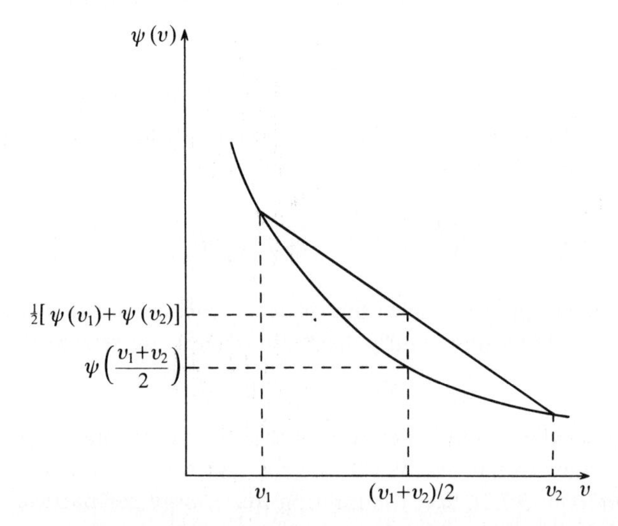

FIG. 3.4 A convex function of v. A chord joining two points always lies above the curve.

The same properties hold of course for concave functions but with the inequalities reversed.

It turns out that stability properties are most easily deduced from corresponding properties of $g(\rho)$. Thus, since g is bounded (eqns (3.7.10), (3.8.5), (3.8.7), and Problem 15) it follows from property 1 that g is a continuous function of ρ, at least for ρ positive and less than the close-packing density ρ_0, and hence from eqn (3.8.8) that ψ is a continuous function of v. Moreover, since the sequence $\{f_k\}$ is monotonic and bounded, we have from eqn (3.8.5) that (Problem 15)

$$g(0) \equiv \lim_{\rho \to 0+} g(\rho) = 0. \tag{3.8.11}$$

Since $g(\rho)$ is continuous and concave, we have by analogy with eqn (3.8.10) that

$$g(\omega_1 \rho_1 + \omega_2 \rho_2) \geqslant \omega_1 g(\rho_1) + \omega_2 g(\rho_2) \tag{3.8.12}$$

for arbitrary non-negative ω_1 and ω_2 such that $\omega_1 + \omega_2 = 1$ and for arbitrary ρ_1 and ρ_2 satisfying $0 \leqslant \rho_i < \rho_0$, $i = 1, 2$.

Setting $\rho_1 = 0$, $\rho_2 = \rho$ and $\omega_2 = \omega$ in eqn (3.8.12) and using eqn (3.8.11), we have

$$g(\omega\rho) \geqslant \omega g(\rho) \qquad \text{for } 0 \leqslant \omega \leqslant 1. \tag{3.8.13}$$

It then follows from eqns (3.8.7) and (3.8.8) that

$$\psi(v) \geqslant \psi(v') \qquad \text{when } v \leqslant v', \tag{3.8.14}$$

i.e. the free energy ψ is a monotonic non-increasing continuous convex function of v.

The pressure given by

$$p = -\frac{\partial \psi}{\partial v} \tag{3.8.15}$$

then exists almost everywhere (property 3) and, from eqn (3.8.14), is non-negative. Moreover, since the derivative of a convex function is non-decreasing, the *pressure is a monotonic non-increasing function of the specific volume v*. Finally, since a monotonic function is differentiable almost everywhere, the isothermal compressibility

$$K_T = -\left(v\frac{\partial p}{\partial v}\right)^{-1} \tag{3.8.16}$$

exists and is non-negative almost everywhere.

To prove that the pressure is a continuous function of v requires more effort, since the derivative of a convex function may have jump discontinuities. Ruelle was the first to prove absolute continuity of the pressure for classical systems but required the additional assumption that $|\phi(r)|$ be bounded. Ruelle's result was extended by Ginibre (1967) and later by others to systems having van Hove-like potentials (see Ruelle 1969).

To establish other stability properties, such as the specific heat at constant volume being non-negative, it suffices to note that the free energy is a concave function of the temperature (see Problem 16).

Similar results to the above hold for magnetic systems (see Problem 12) and for other ensembles. The interested reader is referred to Ruelle (1969) for further details.

3.9 Problems

1. The Laplace transform $\mathscr{L}\{h\}$ of a function h is defined by

$$\mathscr{L}\{h\}=\int_0^{\infty} e^{-sx}h(x)dx$$

and the convolution of two functions f and g is defined by

$$(f*g)(x)=\int_0^x f(x-t)g(t)\,dt.$$

Prove the Laplace convolution formula

$$\mathscr{L}\{f*g\}=\mathscr{L}\{f\}\,\mathscr{L}\{g\}.$$

2. If $f(x)$ is continuous and takes its maximum value at the point $x_0>0$, and the integral

$$I_N=\int_0^{\infty} \exp[Nf(x)]\,dx$$

exists, prove that

$$\lim_{N\to\infty} N^{-1}\ln I_N=f(x_0).$$

3. The Laplace transform of the canonical partition function $Z(l)$ for the Takahashi gas, considered as a function of $l=L-(N-1)a$ is given, from eqn (3.3.9), by

$$\int_0^{\infty} e^{-sl}Z(l)dl=s^{-2}[F(s)]^{N-1}$$

where

$$F(s)=\int_0^a \exp[-sx+\beta\phi(x+a)]dx+s^{-1}e^{-as}.$$

Assuming that, for fixed $v=L/N$,

$$Z(l)\sim\exp[Nf(v,\,T)] \qquad \text{as } N\to\infty$$

show, using the result of Problem 2, that

$$f(v,\,T)=s(v-a)+\ln F(s)$$

with

$$s=\frac{\partial f}{\partial v}=\beta p$$

and p the pressure.

4. Show, explicitly, that the free energy for the Takahashi gas is a monotonic non-increasing continuous convex function of v.

5. Obtain the equation of state for the Takahashi gas with potentials

(a)
$$\phi(x)=\begin{cases}\infty & \text{if } 0\leqslant x<a\\ -c & \text{if } a\leqslant x<2a,\\ 0 & \text{if } x\geqslant 2a\end{cases}$$

(b)
$$\phi(x)=\begin{cases}\infty & \text{if } 0\leqslant x<a\\ c(x-2a) & \text{if } a\leqslant x<2a\\ 0 & \text{if } x\geqslant 2a.\end{cases}$$

6. Show that the isothermal compressibility K_T and the specific heat C_V for the lattice gas are related to the isothermal susceptibility χ_T and specific heat at constant magnetization C_m for the corresponding Ising model by

$$4v^{-2}K_T=\chi_T \qquad \text{and} \qquad v^{-1}C_V=C_m.$$

7. Show that the binary alloy with equal numbers of A and B atoms corresponds to an Ising model with zero magnetization.

8. By differentiating the expression, eqn (3.5.6),

$$\sum_{\{\mu\}}\exp\left[\beta\sum_{i=1}^{N-1}J_i\mu_i\mu_{i+1}\right]=2\prod_{i=1}^{N-1}2\cosh(\beta J_i)$$

with respect to $J_k, J_{k+1},\ldots,J_{k+r-1}$ and setting $J_i=J$ for all $i=1,2,\ldots,N-1$ show that the pair-correlation function for the open chain defined in eqn (3.5.8) is given by

$$\langle\mu_k\mu_{k+r}\rangle=[\tanh(\beta J)]^r \qquad \text{for all } N>1.$$

9. Show that

$$\int\exp(\boldsymbol{\alpha}\cdot\mathbf{S})\mathrm{d}\sigma=2\pi^{n/2}(\|\boldsymbol{\alpha}\|/2)^{1-\frac{n}{2}}I_{\frac{n}{2}-1}(\|\boldsymbol{\alpha}\|)$$

where the integral is taken over the surface $\|\mathbf{S}\|=1$ of the n-dimensional unit sphere and $I_\mu(x)$ is the modified Bessel function of the first kind of order μ defined by eqn (3.5.12).

10. Obtain the expression, eqn (3.5.14), for the pair-correlation function of the open n-vector model in one dimension with nearest-neighbour interactions.

11. Prove Jensen's inequality

$$\langle\exp X\rangle\geqslant\exp\langle X\rangle$$

for the case where the random variable X takes real values and the expectation value $\langle\cdot\rangle$ has the properties $\langle X+Y\rangle=\langle X\rangle+\langle Y\rangle$ and $\langle 1\rangle=1$.

12. The canonical partition function for a translationally invariant one-dimensional Ising model with fixed magnetization

$$m=N^{-1}\sum_{i=1}^{N}\mu_i$$

can be written as

$$Z(N,m,T)=\sum_{\{\mu\}}{}'\exp\left(\beta\sum_{1\leqslant i<j\leqslant N}J(j-i)\mu_i\mu_j\right)$$

where the primed sum is over all configurations

$$\{\mu\}=(\mu_1,\mu_2,\ldots,\mu_N),\ \mu_i=\pm 1 \qquad i=1,2,\ldots,N$$

with fixed m.

(a) Show that

(i) $Z(N, m, T) = Z(N, -m, T)$

and

$$\text{(ii)}\ \langle \mu_i \rangle = [Z(N, m, T)]^{-1} \sum_{\{\mu\}}' \mu_i \exp\left(\beta \sum_{1 \leqslant i < j \leqslant N} J(j-i)\mu_i \mu_j\right)$$

$$= m \qquad \text{for all } i = 1, \ldots, N.$$

(b) By dividing a chain of length $2N$ into two equal chains and fixing the magnetizations per spin of the two halves to be m_1 and m_2, respectively, show using Jensen's inequality, that

$$Z(2N, \tfrac{1}{2}(m_1 + m_2), T) \geqslant Z(N, m_1, T) Z(N, m_2, T) \exp(\beta R_N m_1 m_2)$$

where

$$R_N = \sum_{i=1}^{N} \sum_{j=N+1}^{2N} J(j-i).$$

(c) Assuming that

$$0 \leqslant J(k) \leqslant c/|k|^{1+\varepsilon} \qquad \text{with } c, \varepsilon > 0,$$

show that the limit of the sequence $\{f_n\}$ with

$$f_n(m) = 2^{-n} \ln Z(2^n, m, T) \qquad n = 1, 2, 3, \ldots$$

for fixed m exists and is a concave function of m.

13. If the potential between two particles at positions $\mathbf{r}_i$ and $\mathbf{r}_j$ satisfies

$$\phi(|\mathbf{r}_i - \mathbf{r}_j|) \leqslant C/|\mathbf{r}_i - \mathbf{r}_j|^{d+\varepsilon} \qquad \text{for } |\mathbf{r}_i - \mathbf{r}_j| \geqslant R$$

where d is the dimensionality of the system, and C and ε are positive constants, show that for any two subdomains Ω_1 and Ω_2 of a domain Ω such that the minimum distance between Ω_1 and Ω_2 is at least R,

$$Z(\Omega, N_1 + N_2, T) \geqslant Z(\Omega_1, N_1, T) Z(\Omega_2, N_2, T) \times \exp(-N_1 N_2 \beta C / R^{d+\varepsilon}).$$

14. If $f(v) = v g(1/v)$, $v > 0$ and $g(\rho)$ is a continuous concave function of ρ, i.e.

$$g\left(\sum_{i=1}^{n} \omega_i \rho_i\right) \geqslant \sum_{i=1}^{n} \omega_i g(\rho_i)$$

where $\omega_i \geqslant 0$ and $\sum_{i=1}^{n} \omega_i = 1$, deduce that $f(v)$ is a concave function of v, i.e.

$$f(\tfrac{1}{2}(v_1 + v_2)) \geqslant \tfrac{1}{2}[f(v_1) + f(v_2)].$$

15. Show that the function g defined by eqn (3.8.3) is bounded and that $g(0) = 0$.

16. Show that $\psi(v, T)$ is a concave function of T for fixed v and deduce that for $T_1 < T_2$

$$\psi(v, T_2) \leqslant \psi(v, T_1) + (T_1 - T_2) s(v, T_1)$$

where $s(v, T)$ is the entropy.

3.10 Solutions to problems

1. By definition,

$$\mathscr{L}\{f*g\}=\int_0^\infty \mathrm{e}^{-sx}\left(\int_0^x f(x-t)g(t)\mathrm{d}t\right)\mathrm{d}x$$

$$=\int_0^\infty \mathrm{e}^{-st}g(t)\left(\int_t^\infty \mathrm{e}^{-s(x-t)}f(x-t)\mathrm{d}x\right)\mathrm{d}t.$$

Changing the integration variable x to $y=x-t$ gives the required result.

2. We begin by writing

$$I_N=\exp[Nf(x_0)]J_N$$

where

$$J_N=\int_0^\infty \exp\{N[f(x)-f(x_0)]\}\mathrm{d}x.$$

Since $f(x)-f(x_0)\leqslant 0$, the integral J_N is a decreasing function of N. It follows that

$$N^{-1}\ln I_N-f(x_0)=N^{-1}\ln J_N\leqslant N^{-1}\ln J_1$$

and hence that

$$\lim_{N\to\infty} N^{-1}\ln I_N\leqslant f(x_0). \tag{1}$$

Moreover, given any $\varepsilon>0$, there exists by continuity of f, a $\delta>0$ such that

$$|x-x_0|<\delta \text{ implies } f(x)-f(x_0)>-\varepsilon.$$

The lower bound

$$J_N\geqslant\int_{x_0-\delta}^{x_0+\delta} \exp\{N[f(x)-f(x_0)]\}\mathrm{d}x>2\delta\exp(-N\varepsilon)$$

then shows that

$$\lim_{N\to\infty} N^{-1}\ln I_N\geqslant f(x_0)-\varepsilon. \tag{2}$$

Since ε is arbitrary, the required result follows from the inequalities in eqns (1) and (2).

Comment: The limiting result can be expressed in asymptotic form

$$I_N\sim\exp[Nf(x_0)] \qquad \text{as } N\to\infty.$$

This represents the leading term in an asymptotic expansion for I_N obtained by the so-called *Laplace's method*. The first correction term can be obtained heuristically as follows.

Assuming that f is differentiable and that $f''(x_0)\equiv -A<0$, we can expand f in a Taylor series around x_0 to obtain, since $f'(x_0)=0$, the approximation

$$I_N\sim\int_0^\infty \exp\{N[f(x_0)-\frac{A}{2}(x-x_0)^2]\}\mathrm{d}x.$$

Changing variables to $y=(AN/2)^{1/2}(x-x_0)$, we then have

$$I_N \sim \exp[Nf(x_0)]\,(2/AN)^{1/2}\int_{-x_0(AN/2)^{1/2}}^{\infty}\exp(-y^2)\mathrm{d}y$$

$$\sim \exp[Nf(x_0)](2/AN)^{1/2}\int_{-\infty}^{\infty}\exp(-y^2)\mathrm{d}y$$

$$=(2\pi/AN)^{1/2}\exp[Nf(x_0)].$$

As an example $N!$ can be expressed in the form

$$N!=\int_0^{\infty}x^N\mathrm{e}^{-x}\mathrm{d}x=N^{N+1}I_N$$

with $f(x)=\ln x-x$. In this case $x_0=1$, $A=1$, and Laplace's method gives Stirling's formula

$$N!\sim(2\pi N)^{1/2}N^N\mathrm{e}^{-N}\qquad \text{as } N\to\infty.$$

3. Changing integration variable l to v related through

$$l=Nv-(N-1)a$$

and assuming that for fixed v

$$Z(l)\sim\exp[Nf(v,T)]\qquad \text{as } N\to\infty,$$

we have, from Laplace's method, that

$$s^{-2}[F(s)]^{N-1}=\int_0^{\infty}\exp(-sl)Z(l)\mathrm{d}l$$

$$\sim N\int_{(1-\frac{1}{N})a}^{\infty}\exp\left\{N\left[f(v,T)-sv+s\left(1-\frac{1}{N}\right)a\right]\right\}\mathrm{d}v$$

$$\sim N\exp\{N[f(v_0,T)-sv_0+sa]\}$$

where v_0 maximizes the integrand. That is,

$$\frac{\partial}{\partial v}[f(v,T)-sv]_{v_0}=0$$

and

$$\left.\frac{\partial^2 f}{\partial v^2}\right|_{v_0}\leqslant 0.$$

Since, by definition,

$$\beta p=\frac{\partial f}{\partial v},$$

it follows that

$$f(v,T)=\beta p(v-a)+\ln F(\beta p).$$

4. By considering pressure p as a function of v and T we obtain, from the result of Problem 3, the identity

$$\beta p=\left(\frac{\partial f}{\partial v}\right)_T=\beta p+\beta[(v-a)+F'(\beta p)/F(\beta p)]\left(\frac{\partial p}{\partial v}\right)_T.$$

It follows that either $(\partial p/\partial v)_T = 0$ or

$$v = a - F'(\beta p)/F(\beta p) \tag{1}$$

where $F(s)$ is defined by

$$F(s) = \int_0^a \exp[-sx + \beta\phi(x+a)]\,dx + s^{-1}\exp(-as).$$

It is easily shown that, for $s > 0$,

$$F'(s) < 0 \quad \text{and} \quad F''(s) > 0.$$

That is, $F(s)$ is a continuous strictly decreasing convex function of s for positive s. Moreover, if we recall that the potential $\phi(x)$ vanishes when $x > 2a$, we can write

$$F'(s)/F(s) \equiv -E[x]$$

where the 'expectation value' $E[\cdot]$ is defined by

$$E[g(x)] = [F(s)]^{-1}\int_0^\infty g(x)\exp[-sx + \beta\phi(x+a)]\,dx.$$

It follows easily that

$$\frac{d}{ds}[F'(s)/F(s)] = E[(x - E(x))^2] > 0$$

and hence, from eqn (1), that v is a strictly decreasing continuous function of p and vice versa. In fact, since $\phi(x) = 0$ when $x > 2a$, p is an analytic function of v and T so, as expected, there is no phase transition.

Combining the above results with those of Problem 3 shows that the free energy

$$\psi(v, T) = -kTf(v, T)$$

is a non-increasing continuous convex function of v.

5. (a) When

$$\phi(x) = \begin{cases} \infty & \text{if } 0 \leqslant x < a \\ -c & \text{if } a \leqslant x < 2a \\ 0 & \text{if } x \geqslant 2a \end{cases}$$

$$F(s) = \int_0^a \exp[-sx + \beta\phi(x+a)]\,dx + s^{-1}\exp(-as)$$

$$= s^{-1}\exp(-\beta c) + s^{-1}\exp(-as)[1 - \exp(-\beta c)].$$

The equation of state, given in general by eqn (1) in the solution to Problem 4, is then

$$v = a - (\beta p)^{-1}\left[\frac{1 + (1 + \beta ap e^{-\beta ap}(e^{\beta c} - 1))}{1 + e^{-\beta ap}(e^{\beta c} - 1)}\right].$$

(b) Answer:

$$v = a + \frac{p^2 + ce^{-a\beta(p-c)}(c - a\beta p^2 - 2p + \beta acp)}{p(p-c)[p - ce^{-a\beta(p-c)}]}.$$

6. For the lattice gas with V sites

$$K_T = v^2 \beta V^{-1} \left\langle \left(\sum_{i=1}^{V} t_i - \left\langle \sum_{i=1}^{V} t_i \right\rangle \right)^2 \right\rangle$$

and, for the corresponding magnet,

$$\chi_T = \beta V^{-1} \left\langle \left(\sum_{i=1}^{V} \mu_i - \left\langle \sum_{i=1}^{V} \mu_i \right\rangle \right)^2 \right\rangle.$$

Substituting $t_i = \frac{1}{2}(1-\mu_i)$ gives

$$4v^{-2} K_T = \chi_T.$$

For the lattice gas of N particles on V sites

$$C_V = N^{-1} \left(\frac{\partial E}{\partial T} \right)_V.$$

For the corresponding Ising model

$$C_m = V^{-1} \left(\frac{\partial E}{\partial T} \right)_m.$$

7. For the binary alloy

$$t_i = \begin{cases} 1 & \text{if site } i \text{ is occupied by an A atom} \\ 0 & \text{if site } i \text{ is occupied by a B atom.} \end{cases}$$

For a lattice of V sites the total number of A and B atoms are given, respectively, by

$$N_A = \sum_{i=1}^{V} t_i, \qquad N_B = \sum_{i=1}^{V} (1-t_i).$$

In terms of Ising spin variables $\mu_i = 1 - 2t_i$,

$$N_B - N_A = \sum_{i=1}^{V} \mu_i.$$

8.
$$\frac{\partial^r}{\partial J_{k+r-1} \cdots \partial J_{k+1} \partial J_k} \sum_{\{\mu\}} \exp\left(\beta \sum_{i=1}^{N-1} J_i \mu_i \mu_{i+1} \right)$$

$$= \beta^r \sum_{\{\mu\}} (\mu_k \mu_{k+1})(\mu_{k+1} \mu_{k+2}) \cdots (\mu_{k+r-1} \mu_{k+r}) \exp\left(\beta \sum_{i=1}^{N-1} J_i \mu_i \mu_{i+1} \right)$$

$$= \beta^r \sum_{\{\mu\}} \mu_k \mu_{k+r} \exp\left(\beta \sum_{i=1}^{N-1} J_i \mu_i \mu_{i+1} \right)$$

by virtue of the fact that $\mu_i^2 = 1$.

9. Define

$$J_n(r) = \int \exp(\boldsymbol{\alpha} \cdot \mathbf{S}) \mathrm{d}\sigma_r$$

where $\boldsymbol{\alpha}$ is an n-dimensional vector and the integral is taken over the surface of the n-dimensional sphere $\|\mathbf{S}\| = r$ of radius r.

Clearly,

$$\int_0^\infty \exp(-px)J_n(x^{1/2})\mathrm{d}x/2x^{1/2}$$

$$=\int_0^\infty \exp(-pr^2)J_n(r)\mathrm{d}r$$

$$=\int_{-\infty}^{\infty}\cdots\int \exp\left(-p\sum_{i=1}^{n} S_i^2+\sum_{i=1}^{n}\alpha_i S_i\right)\mathrm{d}S_1\cdots\mathrm{d}S_n$$

$$=(\pi/p)^{n/2}\exp(\alpha^2/4p)$$

where

$$\alpha=\|\boldsymbol{\alpha}\|=\left(\sum_{i=1}^{n}\alpha_i^2\right)^{1/2}.$$

From the Laplace inversion formula we then have, on setting $x=1$, the required result

$$J_n(1)=2\pi^{n/2}(2\pi\mathrm{i})^{-1}\int_{c-\mathrm{i}\infty}^{c+\mathrm{i}\infty}\exp(p+\alpha^2/4p)p^{-(\frac{n}{2}-1)-1}\mathrm{d}p$$

$$=2\pi^{n/2}(\alpha/2)^{1-\frac{n}{2}}I_{\frac{n}{2}-1}(\alpha)\equiv\lambda_1(\alpha)$$

where use has been made of the integral representation, eqn (3.5.12), for the modified Bessel function $I_\mu(z)$.

10. From the integral representation of $I_\mu(x)$ given in Problem 9, it is easy to show that

$$\frac{\mathrm{d}}{\mathrm{d}x}[x^{-\mu}I_\mu(x)]=x^{-\mu}I_{\mu+1}(x).$$

By differentiating the result of Problem 9, with respect to the components of $\boldsymbol{\alpha}$, one then obtains

$$\int \mathbf{S}\exp(\boldsymbol{\alpha}\cdot\mathbf{S})\mathrm{d}\sigma=2\pi^{n/2}(\alpha/2)^{1-\frac{n}{2}}I_{\frac{n}{2}}(\alpha)\hat{\boldsymbol{\alpha}}$$

$$\equiv\lambda_2(\alpha)\hat{\boldsymbol{\alpha}}$$

where $\alpha=\|\boldsymbol{\alpha}\|$ and $\hat{\boldsymbol{\alpha}}$ is the unit vector in the direction of $\boldsymbol{\alpha}$.

In particular, if $\boldsymbol{\alpha}=K\mathbf{S}'$ and $\|\mathbf{S}'\|=1$,

$$\int \mathbf{S}\exp(K\mathbf{S}'\cdot\mathbf{S})\mathrm{d}\sigma=\lambda_2(K)\mathbf{S}'.$$

By integrating successively over the unit spheres $\sigma_1, \sigma_{N-1}, \ldots, \sigma_1$ corresponding to spins $\mathbf{S}_N, \ldots, \mathbf{S}_1$ we then have, for $1<k<k+r<N$,

$$\int\cdots\int \mathbf{S}_k\cdot\mathbf{S}_{k+r}\exp\left(K\sum_{i=1}^{N-1}\mathbf{S}_i\cdot\mathbf{S}_{i+1}\right)\mathrm{d}\sigma_1\cdots\mathrm{d}\sigma_N$$

$$=[\lambda_1(K)]^{k-1}[\lambda_2(K)]^r[\lambda_1(K)]^{N-k}2\pi^{n/2}/\Gamma(n/2) \tag{1}$$

where $\lambda_1(K)$ is defined in Problem 9.

Since the partition function (3.5.13) can be written, from Problem 9, as

$$\int\cdots\int \exp\left(K\sum_{i=1}^{N-1}\mathbf{S}_i\cdot\mathbf{S}_{i+1}\right)\mathrm{d}\sigma_1\cdots\mathrm{d}\sigma_N=[\lambda_1(K)]^{N-1}2\pi^{n/2}/\Gamma\left(\frac{n}{2}\right), \tag{2}$$

the pair-correlation function obtained by dividing eqn (1) by eqn (2) is then given by

$$\langle \mathbf{S}_k \cdot \mathbf{S}_{k+r} \rangle = [\lambda_2(K)/\lambda_1(K)]^r$$
$$= [I_{\frac{n}{2}}(K)/I_{\frac{n}{2}-1}(K)]^r$$

which is the result stated in eqn (3.5.14).

11. For real x, $\exp(x) \geqslant 1+x$. It follows that

$$\langle \exp X \rangle = (\exp\langle X \rangle)\langle \exp(X - \langle X \rangle)\rangle$$
$$\geqslant \exp\langle X \rangle(\langle 1 + X - \langle X \rangle \rangle)$$
$$= \exp\langle X \rangle.$$

12. (a) (i) If all spins are reversed, m is replaced by $-m$ but the exponent in the sum over configurations is unaltered.

(ii) By translational invariance $\langle \mu_i \rangle = c$ is independent of i. Summing over i then gives

$$m = \left\langle \frac{1}{N} \sum_{i=1}^{N} \mu_i \right\rangle = c.$$

(b) Denote by $\{\mu\}$ and $\{\mu'\}$ the configurations of the two halves of the chain and write the interaction energy as

$$E\{\mu\} + E\{\mu'\} + E\{\mu, \mu'\}$$

where $E\{\mu\}$ and $E\{\mu'\}$ denote the interaction energies of the two halves of the chain and $E\{\mu, \mu'\}$ the interaction between them. That is

$$E\{\mu, \mu'\} = -\sum_{i=1}^{N} \sum_{j=N+1}^{2N} J(j-i)\mu_i\mu_j.$$

If we now restrict the magnetizations per spin of the two halves to be m_1 and m_2 and define the probability distribution P by

$$P\{\mu, \mu'\} = [Z(N, m_1, T)Z(N, m_2, T)]^{-1} \exp(-\beta E\{\mu\} - \beta E\{\mu'\}),$$

we have

$$Z\left(2N, \frac{1}{2}(m_1 + m_2), T\right) = Z(N, m_1, T)Z(N, m_2, T) \sum_{\{\mu,\mu'\}} P\{\mu, \mu'\}\exp(-\beta E\{\mu, \mu'\}).$$

The required result follows from Jensen's inequality and the fact that from part (a) (ii)

$$\sum_{\{\mu,\mu'\}} P\{\mu, \mu'\}\mu_i\mu_j = m_1 m_2 \qquad \text{when } 1 \leqslant i \leqslant N \text{ and } N+1 \leqslant j \leqslant 2N.$$

(c) When $0 \leqslant J(k) \leqslant c/|k|^{1+\varepsilon}$, $\quad c, \varepsilon > 0$

$$Z(N, m, T) \leqslant 2^N \exp\left[\beta \sum_{1 \leqslant i < j \leqslant 2N} J(j-i)\right]$$
$$\leqslant 2^N \exp\left[\beta N \sum_{k=1}^{\infty} J(k)\right].$$

It follows that the sequence $\{f_n\}$ is bounded above by

$$f_n(m) = 2^{-n} \ln Z(2^n, m, T)$$
$$\leqslant \ln 2 + \beta \sum_{k=1}^{\infty} J(k).$$

Setting $m_1 = m_2 = m$ in the inequality of part (b) gives

$$Z(2^{n+1}, m, T) \geqslant [Z(2^n, m, T)]^2 \exp(\beta R_{2^n} m^2).$$
$$\geqslant [Z(2^n, m, T)]^2.$$

It follows that the sequence $\{f_n\}$ is monotonically increasing and hence

$$\lim_{n\to\infty} f_n(m) = f(m) \qquad \text{exists.}$$

For arbitrary m_1 and m_2 the inequality in part (b) then gives

$$f\left(\frac{m_1+m_2}{2}\right) = \lim_{N\to\infty} (2N)^{-1} \ln Z\left(2N, \frac{1}{2}(m_1+m_2), T\right)$$
$$\geqslant \frac{1}{2}\left[\lim_{N\to\infty} N^{-1} \ln Z(N, m_1, T) + \lim_{N\to\infty} N^{-1} \ln Z(N, m_2, T)\right]$$
$$+ \lim_{N\to\infty} 2N^{-1} \beta R_N m_1 m_2$$
$$= \frac{1}{2}[f(m_1) + f(m_2)]$$

since

$$(2N)^{-1} R_N \leqslant (2N)^{-1} \sum_{k=1}^{2N} \frac{c}{k^{\varepsilon}}$$

which approaches zero as N approaches infinity.

13. Restrict the domain of integration to $\Omega_1 \cup \Omega_2$ configurations to those having N_1 particles in Ω_1 and N_2 particles in Ω_2. With these restrictions the total interaction energy between particles in Ω_1 and Ω_2 is bounded above by

$$N_1 N_2 C / R^{d+\varepsilon}.$$

The result then follows easily by noting that there are $(N_1+N_2)/N_1!N_2!$ ways of restricting configurations, as indicated above.

14. By definition,

$$f\left(\frac{1}{2}(v_1+v_2)\right) = \frac{1}{2}(v_1+v_2) g\left(\frac{2}{v_1+v_2}\right)$$
$$= \frac{1}{2}(v_1+v_2) g\left[\left(\frac{v_1}{v_1+v_2}\right)\frac{1}{v_1} + \left(\frac{v_2}{v_1+v_2}\right)\frac{1}{v_2}\right]$$
$$\geqslant \frac{1}{2}\left[v_1 g\left(\frac{1}{v_1}\right) + v_2 g\left(\frac{1}{v_2}\right)\right] = \frac{1}{2}[f(v_1) + f(v_2)]$$

where the inequality follows from concavity of g with

$$\omega_1 = \frac{v_1}{v_1 + v_2} \quad \text{and} \quad \omega_2 = \frac{v_2}{v_1 + v_2}.$$

15. The required results follow from the definition

$$g(\rho) = \rho f(v)$$

and the inequality (with $v = \rho^{-1}$),

$$f(v) \leqslant 1 + \beta B - \ln \rho.$$

16. The specific heat at constant volume, given by

$$C_V = -T\left(\frac{\partial^2 \psi}{\partial T^2}\right)_V$$

is non-negative (Problem 7 of Chapter 2). It follows immediately that $\psi(v, T)$ is a concave function of T.

Since tangents to a concave function always lie above the function,

$$\left.\frac{\partial \psi}{\partial T}\right|_{T_1} \geqslant \frac{\psi(v, T_2) - \psi(v, T_1)}{T_2 - T_1} \qquad \text{for } T_1 < T_2.$$

The required inequality follows by noting that the entropy is given by

$$s = \left(\frac{\partial \psi}{\partial T}\right)_v.$$

4

THEORIES OF PHASE TRANSITIONS

4.1 Introduction

The occurrence of phase transitions is a familiar and much studied phenomenon. We are all aware, for example, of the transition from water to steam when water is heated to its boiling point and to ice when it is cooled to its freezing point. These are examples of phase-transition points where water undergoes an abrupt change from one 'phase' to another. Related examples are the condensation of a gas to a liquid under compression at sufficiently low temperatures and the abrupt loss of magnetization when a magnet is heated to its Curie point or critical temperature. In this chapter, we will concern ourselves solely with the gas–liquid and magnetic transitions.

The gas–liquid transition is most easily described by reference to Fig. 4.1. The full curves, called *isotherms*, represent pressure (p) as a function of specific volume (v) for fixed temperature. It will be seen that when the gas is compressed at a fixed temperature T less than the *critical temperature* T_c, the gas reaches a certain specific volume v_g where it condenses at constant pressure to a liquid specific volume v_ℓ. For temperature T greater

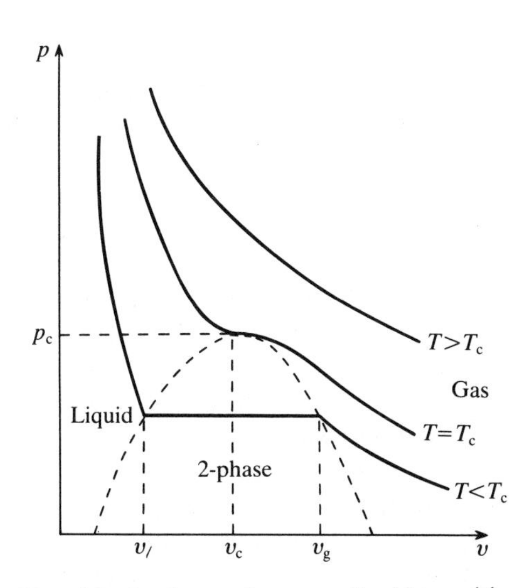

FIG. 4.1 Isotherms for a gas–liquid transition.

than T_c, however, the gas does not condense to a liquid no matter how much it is compressed. The dashed curve in Fig. 4.1 is called the *coexistence curve.* Below this curve we have the *coexistence region* in which the two phases, gas and liquid, coexist.

For the gas–liquid transition, the point (p_c, v_c) on the *critical isotherm*, $T = T_c$, is usually referred to as the critical point. As this point is approached, for example along the *critical isochore*, $v = v_c$, the specific heat and compressibility typically diverge, the latter being observed as critical opalescence caused by large density fluctuations in the neighbourhood of the critical point where the gas can't quite decide whether to remain a gas or condense to a liquid.

Since the gas–liquid transition is typically characterized by abrupt changes in thermodynamic quantities it is natural to associate a phase transition with singular behaviour of the free energy. For finite systems, however, we have seen that the partition function $Z(V, N, T)$, being an integral over a finite domain of an analytic function of temperature $[\exp(-\mathscr{H}/kT)]$ for $T \neq 0$, is an analytic function of $T \neq 0$. In order to give a mathematically precise definition of a phase-transition point we must therefore take the thermodynamic limit, $N, V \to \infty$ with fixed specific volume $v = V/N$. We then arrive at the following definition

Any singular point of the canonical free energy $\psi(v, T)$ given by

$$-\beta\psi(v, T) = \lim_{\substack{N, V \to \infty \\ v = V/N \text{ fixed}}} N^{-1} \ln Z(V, N, T),$$

occurring for real positive T or v is called a phase-transition point.

By singular point we mean any point of non-analyticity of $\psi(v, T)$; or, in other words, any $v = v_c$ or $T = T_c$ around which ψ has no convergent Taylor expansion. It is not necessary, therefore, for any particular thermodynamic quantity obtained by differentiating ψ one or more times with respect to v or T to have a discontinuity or become infinite at a critical point but this is usually what happens.

For magnetic transitions we have a similar situation, as shown in Fig. 4.2 where we have drawn isotherms for various temperatures. In such systems a magnetization M is induced by a magnetic field H with sign depending on the sign (or direction) of the field. For temperatures T below the critical temperature T_c a residual or *spontaneous magnetization* $\pm M_0$ remains when the field H is switched off ($H \to 0\pm$, respectively) whereas when T is larger than T_c there is zero residual magnetization. At $T = T_c$ and $H = 0$, the isothermal susceptibility $(\chi = \partial M/\partial H)$ is typically unbounded and, as T approaches T_c from below, the spontaneous magnetization usually vanishes abruptly, as shown in Fig. 4.3.

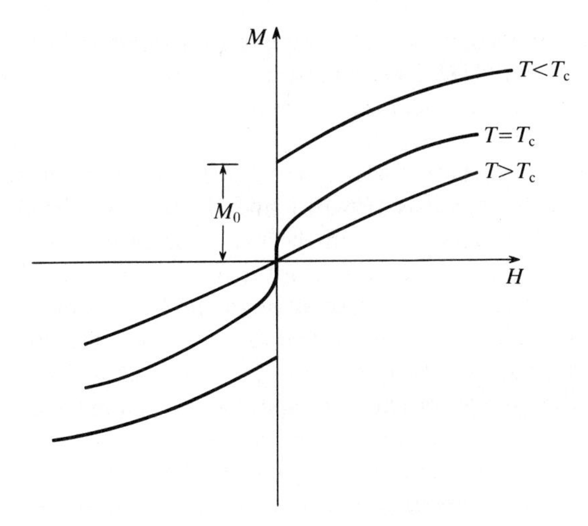

FIG. 4.2 Isotherms for a magnetic transition.

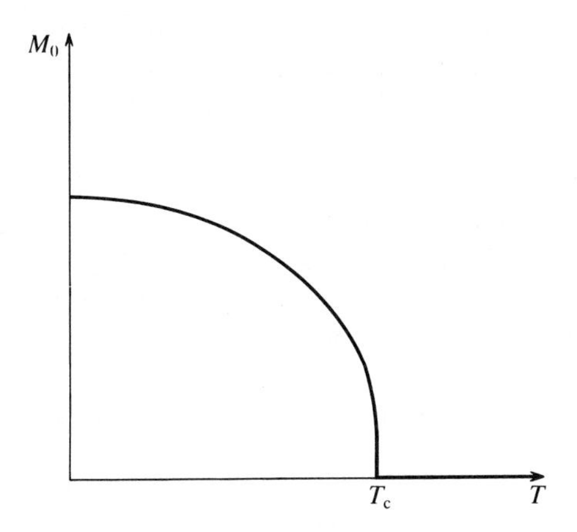

FIG. 4.3 The spontaneous magnetization vanishes abruptly at $T = T_c$ and is zero for $T \geqslant T_c$.

Again, we define a phase-transition point for such systems to be any point of non-analyticity of the canonical free energy $\psi(H, T)$ given by

$$-\beta\psi(H, T) = \lim_{N\to\infty} N^{-1} \ln Z(N, H, T)$$

occurring for real H and/or real positive T, where $Z(H, N, T)$ is the canonical partition function defined for a model spin system, for example, by eqn (3.4.6). We will see later that $\psi(H, T)$ is usually analytic in T and H for $H \neq 0$ so that phase transitions in our strict mathematical sense can only occur in zero field.

As mentioned in Chapter 1, there is a close thermodynamic analogy between fluid and magnetic systems which can be seen when one compares Figs. 4.1 and 4.2 and identifies M with V and H with $-P$. The main difference between the two systems is that in the magnetic case the flat parts of the isotherms lie on top of one another (on the M-axis). It is this feature that in many instances makes the discussion of magnetic systems much simpler than the corresponding fluid system. Notice also that, for the fluid system, the length of the flat parts of the isotherms $v_g - v_\ell$ is positive for $T < T_c$ and like the corresponding *order parameter* or spontaneous magnetization M_0, vanishes abruptly as T approaches T_c from below.

One of the interesting and challenging problems in the theory of phase transitions is to study how quantities such as order parameters behave in the neighbourhood of T_c, and more generally to isolate the essential features of a system that are required to characterize the singular behaviour corresponding to a phase transition. Prior to 1940, in fact, it was commonly felt that such singular behaviour could not be obtained from the partition function or Gibbs formulation alone, the feeling being that something extra was needed for the molecules to know when they should condense to form a liquid or for spins to know when they should become ordered. Thanks to Onsager, however, who showed rigorously in 1944 that the free energy of the two-dimensional Ising model in zero field is singular at a finite temperature, it is now accepted that the Gibbs formulation contains all the necessary ingredients for a rigorous discussion of phase transitions. We will have more to say about this and related models in the following chapter. We turn now to a discussion of the so-called classical theories of phase transitions developed by van der Waals and Curie and Weiss.

4.2 The van der Waals theory of gas–liquid transitions

The celebrated and classical van der Waals equation of state for a gas–liquid phase transition is usually written in the form

$$(v - v_0)\left(p + \frac{\alpha}{2v^2}\right) = kT \tag{4.2.1}$$

where $\rho_0 = v_0^{-1}$ is the close-packing density and α is a positive constant which is usually expressed as the integral of the attractive part of the potential.

This equation has enjoyed considerable success since it first appeared in van der Waals's dissertation of 1873. Away from the critical point it still provides a remarkably good fit to experiment for reasons which will become clear later in this chapter. Close to the critical point however, the theory breaks down and does not provide an adequate fit to modern experimental data.

Although van der Waals's derivation of eqn (4.2.1) predates Gibbs it is possible to give a simple derivation based on the canonical distribution. This, in fact, was first done heuristically by Ornstein in his dissertation of 1908 and is based on the assumption that the attractive part of the potential is long-ranged.

The Ornstein argument in its most elementary form assumes that the gas is composed of hard-sphere particles with pairwise attraction of equal strength between all pairs of particles. That is, if we have N particles contained in a region with volume V, the interaction potential between any two particles separated by a distance r can be written as

$$\phi(r)=q(r)-\frac{\alpha}{V} \qquad (\alpha>0) \tag{4.2.2}$$

where

$$q(r)=\begin{cases}\infty & \text{when } r\leqslant r_0\\ 0 & \text{when } r>r_0\end{cases} \tag{4.2.3}$$

represents the hard-core contribution and, as we will see in a moment, we have included the factor V^{-1} in the equal-strength attractive part of the potential in order to ensure the existence of the thermodynamic limit.

For the gas with potential eqn (4.2.2) the canonical partition function is given by

$$\begin{aligned}Z(V,N,T)&=(N!)^{-1}\int_V\cdots\int\exp\left(-\beta\sum_{1\leqslant i<j\leqslant N}\phi(|\mathbf{r}_i-\mathbf{r}_j|)\right)d\mathbf{r}_1\cdots d\mathbf{r}_N\\ &=\exp(\beta\alpha N(N-1)/2V)Z_{\text{hc}}(V,N)\end{aligned} \tag{4.2.4}$$

where, from (4.2.3),

$$Z_{\text{hc}}(V,N)=(N!)^{-1}\int_V\cdots\int\exp\left[-\beta\sum_{1\leqslant i<j\leqslant N}q(|\mathbf{r}_i-\mathbf{r}_j|)\right]d\mathbf{r}_1\cdots d\mathbf{r}_N \tag{4.2.5}$$

is the partition function for a gas of hard-sphere particles with diameter r_0. As noted in the previous chapter, an exact evaluation of eqn (4.2.5) is an unsolved problem except in one dimension where, for the so-called Tonks gas, we have from eqn (3.2.9)

$$Z_{\text{hc}}(V,N)=(N!)^{-1}[V-(N-1)v_0]^N \tag{4.2.6}$$

where $v_0=r_0$ is now the length of the hard rods. Combining eqns (4.2.4) and (4.2.6) we obtain, in the thermodynamic limit ($v>v_0$),

$$\begin{aligned}-\beta\psi(v,T)&=\lim_{\substack{N,V\to\infty\\ v=V/N\text{ fixed}}} N^{-1}\ln Z(V,N,T)\\ &=1+\beta\alpha/2v+\ln(v-v_0).\end{aligned} \tag{4.2.7}$$

It then follows that

$$p = -\frac{\partial \psi}{\partial v} = kT(v - v_0)^{-1} - \alpha/2v^2 \tag{4.2.8}$$

which is equivalent to the van der Waals equation of state eqn (4.2.1).

Although eqn (4.2.8) is only strictly valid in one dimension (for the potential eqn (4.2.2)), it provides a reasonable approximation in higher dimensions if v_0 is, loosely speaking, interpreted as the 'volume' of a single particle. More correctly, one should write, in place of eqn (4.2.8),

$$p = p_{\text{hc}} - \alpha/2v^2 \tag{4.2.9}$$

where p_{hc} is the pressure of the hard-core system derived from $Z_{\text{hc}}(V, N)$ which, it will be noted, is independent of temperature.

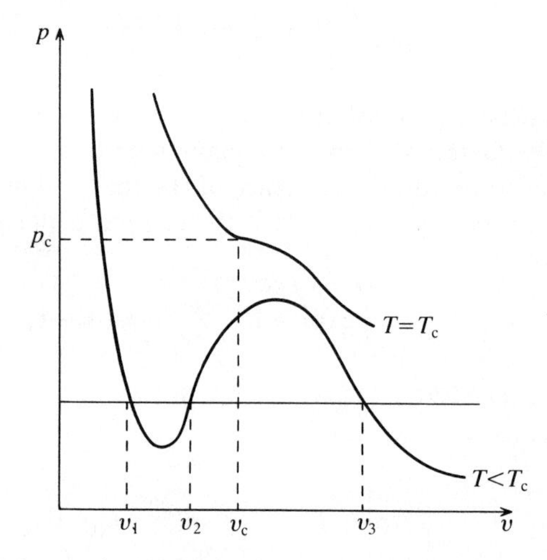

FIG. 4.4 van der Waals isotherms obtained from eqn (4.2.8).

Typical isotherms derived from the van der Waals equation of state (4.2.8) are shown in Fig. 4.4. Rewriting eqn (4.2.8) in the form

$$v^3 - \left(v_0 + \frac{kT}{p}\right)v^2 + \frac{\alpha}{2p}v - \frac{\alpha}{2p}v_0 = 0 \tag{4.2.10}$$

shows that, for fixed p and T, the van der Waals equation of state reduces to a cubic equation for v which in general has three solutions (for $T < T_c$) as shown in Fig. 4.4. As T approaches T_c from below, these three solutions coalesce to the critical value v_c which means that the left-hand side of eqn (4.2.10) when

$p=p_c$ and $T=T_c$ is equivalent to

$$(v-v_c)^3=v^3-3v_cv^2+3v_c^2v-v_c^3. \tag{4.2.11}$$

Equating coefficients and solving the resulting equations then gives the critical point values

$$T_c=4\alpha/27v_0k, \qquad p_c=\alpha/54v_0^2, \qquad v_c=3v_0. \tag{4.2.12}$$

Alternatively, one can derive these critical values by noting that the critical point is a point of inflection on the critical isotherm $T=T_c$.

An interesting feature of the van der Waals equation of state is that it has the universal form (see Problem 1)

$$\tilde{p}=8\tilde{T}(3\tilde{v}-1)^{-1}-3\tilde{v}^{-2} \tag{4.2.13}$$

when expressed in terms of the scaled quantities

$$\tilde{p}=p/p_c, \qquad \tilde{T}=T/T_c, \qquad \tilde{v}=v/v_c. \tag{4.2.14}$$

This universal form is usually referred to as the *law of corresponding states* and is well approximated in nature, away from the critical point.

Another feature of the van der Waals equation of state is that it is unphysical, in the sense that for $T<T_c$ there is a violation of the stability condition $\partial p/\partial v \leqslant 0$, which, as shown in Section 3.8, is a necessary consequence of the Gibbs prescription. This violation of thermodynamic stability was recognized almost immediately by Maxwell (1874) who proposed that the van der Waals 'wiggle' be replaced by the constant pressure portion of each isotherm for $T<T_c$ as shown in Fig. 4.5, in such a way that the area of

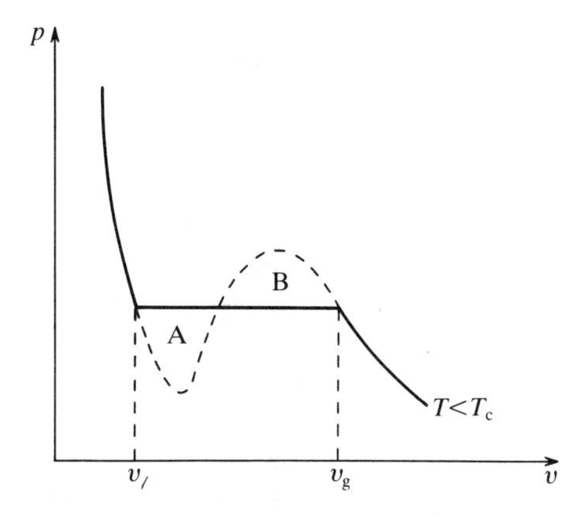

FIG. 4.5 Maxwell construction on the van der Waals isotherm is such that regions A and B have equal areas.

the region A is equal to the area of the region B. That is,

$$p_1(v_g - v_\ell) = \int_{v_1}^{v_g} p \, dv \tag{4.2.15}$$

where the integral is taken along the van der Waals isotherm. This is known as the *Maxwell construction.*

In terms of the free energy $\psi(v, T)$ we have that $p = -\partial\psi/\partial v$ and hence, from eqn (4.2.15), that in the coexistence region for fixed T,

$$p_\ell = -\frac{\psi(v_g, T) - \psi(v_\ell, T)}{v_g - v_\ell} = p_g. \tag{4.2.16}$$

It follows that in this region the Maxwell construction is equivalent to the *double-tangent construction* on the free energy as shown in Fig. 4.6.

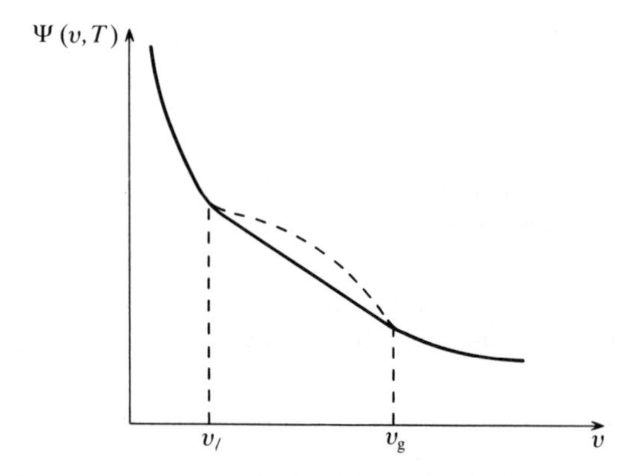

FIG. 4.6 The free energy $\psi(v, T)$ obtained from the double-tangent construction is the convex envelope of the van der Waals free energy.

It will be noted that the modified free energy is a convex function of v as required by thermodynamic stability. In fact the new free energy is the maximal convex function which is not greater than the original van der Waals free energy. Mathematically, such a function is called the *convex envelope* and is denoted by $CE\{f(x)\}$. We will see in Section 4.4 that the modified van der Waals free energy, $\psi_{\text{vdw}}(V, T)$, written quite generally (from eqn (4.2.4)) as

$$\psi_{\text{vdw}}(v, T) = CE\{\psi^{\text{hc}}(v) - \alpha/2v\}, \tag{4.2.17}$$

with ψ^{hc} denoting the free energy of the hard-core reference system, can be derived rigorously from a suitable limiting process corresponding to an infinitely weak long-range potential.

From the point of view of phase transitions one is interested in the nature of the critical point and, more particularly, in the behaviour of certain quantities in the neighbourhood of this point, such as the compressibility and specific heat, as well as in the shape of the coexistence curve and critical isotherm.

For simplicity, we analyse the universal form of the van der Waals equation of state (4.2.13) and, for convenience, denote the scaled variables by p, T, and v. In this form the critical point occurs at $p_c = T_c = v_c = 1$.

Substituting (4.2.13) into (4.2.15) we obtain the convexity condition

$$p_\ell (v_g - v_\ell) = \frac{8T}{3} \ln\left(\frac{3v_g - 1}{3v_\ell - 1}\right) + 3\left(\frac{1}{v_g} - \frac{1}{v_\ell}\right) \tag{4.2.18}$$

for temperatures below the critical point. Close to the critical point, this condition becomes

$$p_\ell (\varepsilon_g + \varepsilon_\ell) = (4T - 3)(\varepsilon_g + \varepsilon_\ell) + O(\varepsilon^2) \tag{4.2.19}$$

where

$$\varepsilon_g = v_g - 1 \qquad \text{and} \qquad \varepsilon_\ell = 1 - v_\ell . \tag{4.2.20}$$

That is, to leading order,

$$p_\ell = p_g \sim 1 - 4(1 - T) \qquad \text{as } T \to 1- . \tag{4.2.21}$$

In general, when we set $v = 1 + \varepsilon$ in eqn (4.2.13) and expand in powers of ε, we obtain

$$p = 1 + (T - 1)(4 - 6\varepsilon + 9\varepsilon^2) - \left(\frac{27T}{2} - 12\right)\varepsilon^3 + O(\varepsilon^4). \tag{4.2.22}$$

In particular, when we substitute (4.2.21) for p in eqn (4.2.22) and solve for ε, we obtain $\varepsilon = \pm 2(1 - T)^{1/2}$ as $T \to 1-$. Comparison with eqn (4.2.20) then gives, to leading order,

$$v_g \sim 1 + 2(1 - T)^{1/2} \qquad \text{and} \qquad v_\ell \sim 1 - 2(1 - T)^{1/2}. \tag{4.2.23}$$

It follows that

$$v_g - v_\ell \sim 4(1 - T)^{1/2} \qquad \text{as } T \to 1- \tag{4.2.24}$$

and hence that in the neighbourhood of the critical point the coexistence curve is parabolic.

Returning to eqn (4.2.22) and setting $T = 1$ shows that in the neighbourhood of the critical point the critical isotherm has the cubic form

$$p \sim 1 - \tfrac{3}{2}(v - 1)^3 \qquad \text{as } v \to 1. \tag{4.2.25}$$

Also, from the equation of state eqn (4.2.13), we have that

$$\frac{\partial p}{\partial v} = -24T(3v - 1)^2 + 6v^{-3} \tag{4.2.26}$$

provided we are outside the coexistence region. In particular, on the *critical isochore*, $v=1$, we easily obtain

$$K_T^{\mathrm{isoc}} = -\left(v\frac{\partial p}{\partial v}\right)^{-1}_{v=1} = \frac{1}{6}(T-1)^{-1} \qquad \text{for } T>1 \tag{4.2.27}$$

and on the coexistence curve $(v \sim 1 \pm 2(1-T)^{1/2})$

$$K_T^{\mathrm{coex}} \sim \tfrac{1}{12}(1-T)^{-1} \qquad \text{as } T\to 1-. \tag{4.2.28}$$

The isothermal compressibility (K_T), therefore, has a simple pole singularity in the temperature at the critical point.

Finally, to investigate the behaviour of the specific heat (at constant volume)

$$C_v = -T\frac{\partial^2 \psi}{\partial T^2} \tag{4.2.29}$$

in the neighbourhood of the critical point we note first of all, from eqn (4.2.7), that, above the critical temperature,

$$C_v = 0 \qquad \text{for all } T>T_c. \tag{4.2.30}$$

(Actually, since we have ignored the kinetic-energy contribution (3.1.4) to the partition function, the right-hand side of eqn (4.2.30) should really be replaced by the specific heat value of $3k/2$ for the ideal gas.)

Below T_c in the coexistence region $v_\ell < v < v_g$, the convex envelope construction eqn (4.2.17) amounts to writing the free energy as

$$\psi(v, T) = \psi_w(v_\ell, T) + (v - v_\ell)\left[\frac{\psi_w(v_g, T) - \psi_w(v_\ell, T)}{v_g - v_\ell}\right] \tag{4.2.31}$$

where $\psi_w(v, T)$ denotes the van der Waals free energy (4.2.7) and use has been made of eqn (4.2.16).

If we set $v = v_c$ and expand $\psi_w(v_g, T)$ and $\psi_w(v_\ell, T)$ in power series around v_c with T fixed, we obtain, to leading order, in unscaled variables

$$\psi(v_c, T) = \psi_w(v_c, T) - \frac{\varepsilon^2}{2!}\frac{\partial p_w}{\partial v}\bigg|_{v_c, T} - \frac{\varepsilon^4}{4!}\frac{\partial^3 p_w}{\partial v^3}\bigg|_{v_c, T} + \cdots \tag{4.2.32}$$

where $\varepsilon = \frac{1}{2}(v_g - v_\ell)$ and p_w is given by the van der Waals expression (4.2.8). It is then a straightforward but somewhat tedious matter to show that on the critical isochore (Problem 2)

$$\lim_{T\to T_c-} C_{v_c} = 9k/2. \tag{4.2.33}$$

It thus follows that the specific heat has a jump discontinuity at the critical point.

4.3 The Curie–Weiss theory of magnetic transitions

The simplest model of a magnetic system is the Ising model with pairwise interaction. For such a system composed of N spins $\mu_i = \pm 1$, $i = 1, 2, \ldots, N$ with coupling constants J_{ij} between spins i and j and external magnetic field H, the interaction energy in a given configuration $\{\mu\} = (\mu_1, \mu_2, \ldots, \mu_N)$ is given by

$$E\{\mu\} = -\sum_{1 \leq i < j \leq N} J_{ij}\mu_i\mu_j - H\sum_{i=1}^{N} \mu_i. \tag{4.3.1}$$

The problem is to calculate the partition function

$$Z_N(\beta, H) = \sum_{\{\mu\}} \exp(-\beta E\{\mu\}) \tag{4.3.2}$$

where the sum over $\{\mu\}$ is over all possible 2^N configurations of spins. Thermodynamic properties are then obtained in the usual way from the limiting free energy $\psi(\beta, H)$ defined by

$$-\beta\psi(\beta, H) = \lim_{N \to \infty} N^{-1} \ln Z_N(\beta, H). \tag{4.3.3}$$

In the spirit of Ornstein's observation for the gas, the Curie–Weiss theory for magnetic systems is based on the assumption that when the range of the interaction is very long, each spin on average interacts with a *mean field* produced by all the other spins. This means that for most configurations each spin has an effective constant interaction with all the other spins. In its simplest form, then, the Curie–Weiss theory, or mean-field theory as it is often called, assumes that the coupling constants J_{ij} in eqn (4.3.1) are independent of i and j and hence that the interaction energy has the form

$$E\{\mu\} = -\frac{J}{N}\sum_{1 \leq i < j \leq N} \mu_i\mu_j - H\sum_{i=1}^{N} \mu_i \tag{4.3.4}$$

where, as in Ornstein's case for the gas, the factor N^{-1} in eqn (4.3.4) is required for the thermodynamic limit eqn (4.3.3) to exist. The model having interaction energy (4.3.4) is usually referred to as the Curie–Weiss model. For the present discussion we will assume that $J > 0$, i.e. the system is ferromagnetic.

In order to evaluate the partition function eqn (4.3.2) for the Curie–Weiss model we first use the fact that $\mu_i^2 = 1$ to express the interaction energy eqn (4.3.4) in symmetric form

$$E\{\mu\} = -\frac{J}{2N}\left(\sum_{i=1}^{N} \mu_i\right)^2 + \frac{J}{2} - H\sum_{i=1}^{N} \mu_i. \tag{4.3.5}$$

The partition function eqn (4.3.2) can then be expressed in the form

$$Z_N(\beta, H) = \exp(-K/2) \sum_{\{\mu\}} \exp\left[\frac{K}{2N}\left(\sum_{i=1}^{N} \mu_i\right)^2 + B \sum_{i=1}^{N} \mu_i\right] \tag{4.3.6}$$

where

$$K = \beta J \quad \text{and} \quad B = \beta H. \tag{4.3.7}$$

We now make use of the identity

$$\exp(a^2/2) = (2\pi)^{-1/2} \int_{-\infty}^{\infty} \exp(-x^2/2 + ax)\mathrm{d}x \tag{4.3.8}$$

which is easily established by completing the square in the exponent on the right-hand side and using the fact that

$$(2\pi)^{-1/2} \int_{-\infty}^{\infty} \exp(-y^2/2)\mathrm{d}y = 1. \tag{4.3.9}$$

Choosing

$$a = (K/N)^{1/2} \sum_{i=1}^{N} \mu_i \tag{4.3.10}$$

in eqn (4.3.8) then enables us to express the partition function (4.3.6) in the form

$$Z_N(\beta, H) = \exp(-K/2) \sum_{\{\mu\}} (2\pi)^{-1/2} \int_{-\infty}^{\infty} \exp\left[-x^2/2 + \left(x\left(\frac{K}{N}\right)^{\frac{1}{2}} + B\right) \sum_{i=1}^{N} \mu_i\right]\mathrm{d}x. \tag{4.3.11}$$

The spins in eqn (4.3.11) are now decoupled so that the sum over spin configurations can be performed to give

$$Z_N(\beta, H) = \exp(-K/2)(2\pi)^{-1/2} \int_{-\infty}^{\infty} \exp(-x^2/2)\left[2\cosh\left(x\left(\frac{K}{N}\right)^{\frac{1}{2}} + B\right)\right]^N \mathrm{d}x \tag{4.3.12}$$

where use has been made of the fact that for each spin

$$\begin{aligned} \sum_{\mu_i = \pm 1} \exp(L\mu_i) &= \exp(L) + \exp(-L) \\ &= 2\cosh L. \end{aligned} \tag{4.3.13}$$

Changing variables to $m = x(KN)^{-1/2}$ we obtain

$$Z_N(\beta, H) = (KN/2\pi)^{1/2} \exp(-K/2) \int_{-\infty}^{\infty} \exp[Nf(m)]\mathrm{d}m \tag{4.3.14}$$

where

$$f(m) = -Km^2/2 + \ln 2\cosh(Km + B). \tag{4.3.15}$$

For large N the integral in eqn (4.3.14) can be evaluated asymptotically by Laplace's method (see Problem 2 of Chapter 3) so that in the thermodynamic limit we obtain, from eqns (4.3.3) and (4.3.14),

$$-\beta\psi(\beta, H) = \max_{|m|<\infty} f(m). \tag{4.3.16}$$

Differentiating $f(m)$ with respect to m, we easily find that the equation determining the maximum in eqn (4.3.16) is the classical mean-field equation

$$m = \tanh(Km + B). \tag{4.3.17}$$

Moreover, if we substitute the maximizing solution $m(\beta, H)$ into eqn (4.3.16) and differentiate with respect to B, we find that

$$m(\beta, H) = \frac{\partial}{\partial B}(-\beta\psi) \tag{4.3.18}$$

is precisely the magnetization of the system. In particular, if we take the limit $H \to 0+$, the non-negative solution of the equation

$$m_0 = \tanh(Km_0) \tag{4.3.19}$$

is the spontaneous magnetization of the system.

The form of m_0 is most easily seen by reference to Fig. 4.7. Thus, when the slope K of $\tanh(Kx)$ at the origin exceeds unity, we have three solutions 0, $\pm m_0$ of eqn (4.3.19), whereas, when K is less than unity, there is only one solution, 0, corresponding to zero spontaneous magnetization for all

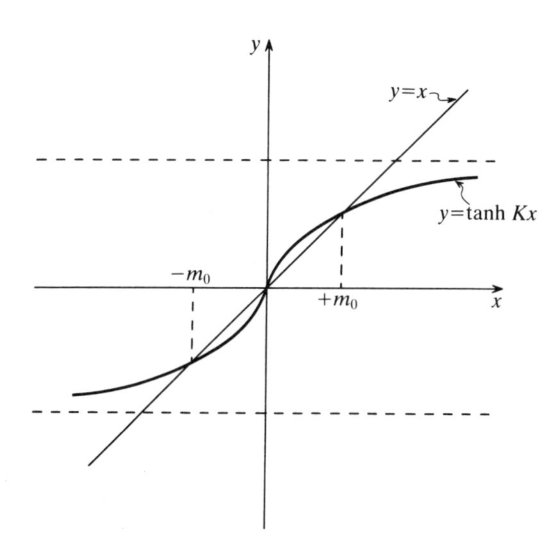

FIG. 4.7 Graphical solution of eqn (4.3.19).

temperatures T exceeding the critical temperature T_c given by

$$K_c = \frac{J}{kT_c} = 1. \tag{4.3.20}$$

An elementary calculation shows that when T is less than T_c, the non-zero solution of eqn (4.3.19) maximizes $f(m)$ so that the spontaneous magnetization has the form shown in Fig. 4.3.

Just below the critical temperature m_0 is small so that $\tanh(Km_0)$ in eqn (4.3.19) can be expanded to give

$$m_0 = Km_0 - (Km_0)^3/3 + O(m_0^5) \qquad \text{as } T \to T_c- \tag{4.3.21}$$

from which one easily obtains

$$m_0 \sim \left[3\left(1 - \frac{T}{T_c}\right)\right]^{1/2} \qquad \text{as } T \to T_c-. \tag{4.3.22}$$

It will be noted from eqns (4.2.24) and (4.3.22) that in the neighbourhood of their respective critical points, $v_g - v_\ell$ for the van der Waals gas and m_0 for the Curie–Weiss magnet have the same asymptotic form or *critical exponent* 1/2 as $T \to T_c-$. In fact, as we will see in a moment, the van der Waals gas and the Curie–Weiss magnet have identical critical properties.

The form of the critical isotherm in the Curie–Weiss theory is obtained from eqn (4.3.17) by setting $T = T_c$ (i.e. $K = 1$ and $B = \beta_c H$) and again expanding the tanh function. It follows that at $T = T_c$ and $H \to 0$,

$$m = (m + \beta_c H) - (m + \beta_c H)^3/3 + O(m + \beta_c H)^5 \tag{4.3.23}$$

from which one deduces that

$$H \sim (3J)^{-1} m^3 \qquad \text{at } T = T_c \text{ and } H \to 0 \tag{4.3.24}$$

which is to be compared with eqn (4.2.25) for the van der Waals gas.

The isothermal susceptibility defined by

$$\chi = \partial m / \partial H \tag{4.3.25}$$

is the magnetic analogue of the isothermal compressibility. From eqn (4.3.17) it is easy to show that

$$\chi = \beta(1 - m^2)[1 - K(1 - m^2)]^{-1}. \tag{4.3.26}$$

In zero field the spontaneous magnetization is zero when $T > T_c$ and hence, from eqns (4.3.20) and (4.3.26),

$$\chi_0 = \lim_{H \to 0+} \chi = \beta\left(1 - \frac{T_c}{T}\right)^{-1} \qquad \text{for } T > T_c. \tag{4.3.27}$$

This result is to be compared with eqn (4.2.27) for the van der Waals expression for the compressibility.

Below the critical point, use of eqn (4.3.22) shows that

$$\chi_0 \sim \frac{\beta_c}{2}\left(\frac{T_c}{T}-1\right)^{-1} \qquad \text{as } T \to T_c - 1 \tag{4.3.28}$$

which is to be compared with eqn (4.2.28).

Finally, the zero-field free energy $\psi_0 = \psi(\beta, 0)$ from eqns (4.3.15) and (4.3.16) is given by

$$-\beta\psi_0 = \begin{cases} \ln 2 & \text{when } T \geqslant T_c \\ -Km_0^2/2 + \ln(2\cosh Km_0) & \text{when } T < T_c. \end{cases} \tag{4.3.29}$$

It then easily follows that the zero-field specific heat is given by

$$C_0 = -T\frac{\partial^2 \psi_0}{\partial T^2} = \begin{cases} 0 & \text{when T} \geqslant T_c \\ -\dfrac{J\,\mathrm{d}m_0^2}{2\,\mathrm{d}T} & \text{when } T < T_c \end{cases} \tag{4.3.30}$$

which has a jump discontinuity at T_c. The magnitude of the jump discontinuity is easily found from eqns (4.3.22) and (4.3.30) to be $3k/2$ (see Problem 3).

4.4 Validity of the classical theories

As we remarked in Section 4.2, the van der Waals theory with Maxwell equal area or convex envelope construction is valid for weak long-ranged attractive potentials. In the extreme case of equal strength attraction between all pairs of particles we obtained the unphysical van der Waals wiggle which is a direct consequence of allowing the potential eqn (4.2.2) to depend on the size of the system. It is clear then that one must exercise more care in proceeding to a long-range limit.

The first rigorous derivation of the van der Waals equation of state *with Maxwell construction* was given by Kac, Uhlenbeck, and Hemmer, who, in a series of three papers (Kac *et al.* 1963; Uhlenbeck *et al.* 1963; Hemmer *et al.* 1964), considered a one-dimensional gas of hard rods interacting with an attractive potential of the form

$$\phi(x) = -\gamma\alpha\exp(-\gamma|x|). \tag{4.4.1}$$

What Kac, Uhlenbeck, and Hemmer were able to show was that, if one takes the limit $\gamma \to 0+$ *after* the thermodynamic limit, one rigorously obtains the van der Waals equation of state *with Maxwell construction.* It is to be noted that when γ is small the potential $\phi(x)$ is weak and flat so that in a sense γ^{-1} is a measure of the range of interaction.

Shortly after the work of Kac, Uhlenbeck, and Hemmer, van Kampen (1964) extended Ornstein's original heuristic argument which led Lebowitz and Penrose (1966) to a more general rigorous discussion of the validity of the van der Waals theory.

Specifically, Lebowitz and Penrose considered a gas of hard-sphere particles in d dimensions with an attractive pair potential of the form

$$\phi(\mathbf{r}) = -\gamma^d K(\gamma|\mathbf{r}|). \tag{4.4.2}$$

Assuming only that K is non-negative, bounded, and Riemann-integrable over every bounded region of d-dimensional space, they were able to obtain the convex envelope result (4.2.17) by taking the limit $\gamma \to 0+$ after the thermodynamic limit where

$$\alpha = -\int \phi(\mathbf{r})\mathrm{d}\mathbf{r} > 0 \tag{4.4.3}$$

it will be noted, from eqn (4.4.2), is independent of γ.

Similar results have been obtained for spin systems (see, for example, Baker 1961; Kac and Helfand 1963; Siegert and Vezzetti 1968; Thompson and Silver 1973; Pearce and Thompson 1978) by taking coupling constants to be so-called *Kac potentials* of the form given in eqn (4.4.2) and then taking the limit $\gamma \to 0+$ after the thermodynamic limit. In this way one rigorously recovers the classical mean-field theories for spin systems.

To give some idea of how one goes about proving such results we present here a simplified version of the Lebowitz–Penrose argument which has been specially adapted to treat the one-dimensional Ising model with Kac potential. The main advantage of treating the magnet rather than the gas is that for the magnet there is no convex envelope construction to contend with.

Consider then a one-dimensional Ising model with interaction energy

$$E\{\mu\} = -\gamma \sum_{1 \leq i < j \leq N} K(\gamma|i-j|)\,\mu_i\mu_j - H\sum_{i=1}^{N} \mu_i. \tag{4.4.4}$$

We will assume throughout that K is non-negative and Riemann-integrable over every bounded interval of the real line. The problem is to evaluate the partition function

$$Z_N(\beta, H, \gamma) = \sum_{\{\mu\}} \exp(-\beta E\{\mu\}) \tag{4.4.5}$$

asymptotically for small γ and large N.

Following Lebowitz and Penrose, we obtain upper and lower bounds on Z_N and show that, when the limit $\gamma \to 0+$ is taken after the thermodynamic limit, the two bounds coalesce to give the result

$$\begin{aligned} -\beta\psi(\beta, H) &= \lim_{\gamma \to 0+} \lim_{N \to \infty} N^{-1} \ln Z_N(\beta, H, \gamma) \\ &= \max_{|m| < \infty} \{-Km^2/2 + \ln 2\cosh(Km + \beta H)\} \end{aligned} \tag{4.4.6}$$

which is precisely the Curie–Weiss expression (4.3.16) where now

$$K=\beta\int_{-\infty}^{\infty} K(x)\mathrm{d}x. \tag{4.4.7}$$

We begin by writing the interaction energy eqn (4.4.4) in symmetric form

$$E\{\mu\}=-\frac{\gamma}{2}\sum_{i,j=1}^{N} K(\gamma|i-j|)(1+\mu_i\mu_j)+\frac{\gamma}{2}\sum_{i,j=1}^{N} K(\gamma|i-j|)-H\sum_{i=1}^{N}\mu_i \tag{4.4.8}$$

and in such a way that all terms in the first sum are non-negative (since $K\geqslant 0$ and $\mu_i\mu_j=\pm 1$). Also, for simplicity, we assume here and henceforth that $K(0)=0$.

In order to obtain an upper bound on the partition function we divide the chain of N sites into L strips $\omega_1, \omega_2, \ldots, \omega_L$ each containing s sites (so that $N=Ls$) as shown in Fig. 4.8. We now define

$$Z_N(\beta, H, \gamma; m_1, m_2, \ldots, m_L)=\sum_{\{\mu\}}{}' \exp(-\beta E\{\mu\}) \tag{4.4.9}$$

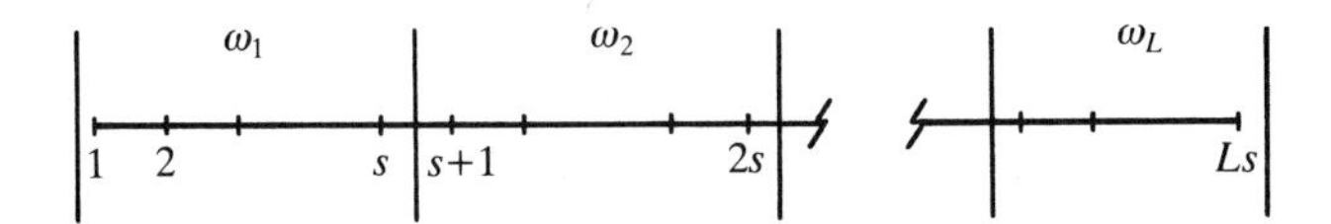

FIG. 4.8 Partitioning of a chain of $N=Ls$ sites into L strips $\omega_1, \omega_2, \ldots, \omega_L$, each containing s sites.

where the sum over $\{\mu\}$ is restricted to configurations with fixed magnetization per site

$$m_k=s^{-1}\sum_{i\in\omega_k}\mu_i \tag{4.4.10}$$

in strip ω_k. Since each m_k can take on only $2s+1$ values it then follows that

$$Z_N(\beta, N, \gamma)=\sum_{m_1,\ldots,m_L} Z_N(\beta, H, \gamma; m_1, m_2, \ldots, m_L)$$

$$\leqslant(2s+1)^L \max_{\{m_k, |m_k|\leqslant 1\}} Z_N(\beta, H, \gamma; m_1, m_2, \ldots, m_L). \tag{4.4.11}$$

Now from eqn (4.4.10) we have

$$\sum_{\substack{i\in\omega_k\\ j\in\omega_l}}(1+\mu_i\mu_j)=s^2(1+m_k m_l). \tag{4.4.12}$$

We then obtain, from eqns (4.4.8) and (4.4.12), the inequality

$$-\beta E\{\mu\} \leqslant \frac{1}{2}\beta\gamma s^2 \sum_{k,l=1}^{L} K^+(|k-l|)(1+m_k m_l) - \frac{1}{2}\beta\gamma \sum_{i,j=1}^{N} K(\gamma|i-j|) + \beta H s \sum_{k=1}^{L} m_k \tag{4.4.13}$$

where

$$K^+(|k-l|) = \max_{\substack{i\in\omega_k \\ j\in\omega_l}} K(\gamma|i-j|). \tag{4.4.14}$$

Using the elementary inequality

$$m_k m_l \leqslant \tfrac{1}{2}(m_k^2 + m_l^2) \tag{4.4.15}$$

it then follows straightforwardly from eqn (4.4.13) that

$$-\beta E\{\mu\} \leqslant \beta s \sum_{k=1}^{L} E^+(m_k) \tag{4.4.16}$$

where, since $N = Ls$,

$$E^+(m) = (1+m^2)\alpha^+ + Hm - \alpha_N, \tag{4.4.17}$$

$$\alpha^+ = \gamma s \sum_{k=0}^{\infty} K^+(k) \tag{4.4.18}$$

and

$$\alpha_N = (2N)^{-1}\gamma \sum_{i,j=1}^{N} K(\gamma|i-j|). \tag{4.4.19}$$

Finally, since there are $s!/d!(s-d)!$ ways of arranging d down spins ($\mu = -1$) among s sites, it follows that there are

$$A(m_k) = s! \Bigg/ \left[\frac{s}{2}(1+m_k)\right]! \left[\frac{s}{2}(1-m_k)\right]! \tag{4.4.20}$$

configurations satisfying the restriction eqn (4.4.10). Combining eqns (4.4.9), (4.4.11), (4.4.16), and (4.4.20) then yields the upper bound

$$Z_N(\beta, H, \gamma) \leqslant (2s+1)^L \max_{\{m_k, |m_k|\leqslant 1\}} \prod_{k=1}^{L} A(m_k) \exp\left(\beta \sum_{k=1}^{L} E^+(m_k)\right)$$

$$= (2s+1)^L \exp\left[N \max_{|m|<1} f_N^+(m)\right] \tag{4.4.21}$$

where, from eqn (4.4.17),

$$f_N^+(m) = \beta[\alpha^+ m^2 + Hm + (\alpha^+ - \alpha_N)] + s^{-1}\ln A(m). \tag{4.4.22}$$

To obtain a lower bound for the partition function we note, from eqn (4.4.11), that for *any* $m_1, m_2, \ldots, m_L$,

$$Z_N(\beta, H, \gamma) \geqslant Z_N(\beta, H, \gamma;\, m_1, m_2, \ldots, m_L). \tag{4.4.23}$$

In particular, if we choose $m_i = m$, $i = 1, 2, \ldots, L$, where m maximizes $f^+(m)$ in eqn (4.4.22) we have, in place of eqn (4.4.13), the inequality

$$-\beta E\{\mu\} \geqslant \tfrac{1}{2}\beta\gamma s^2 \sum_{k,l=1}^{L} K^-(|k-l|)(1+m^2) - \tfrac{1}{2}\beta\gamma \sum_{i,j=1}^{N} K(\gamma|i-j|) + N\beta H m \tag{4.4.24}$$

where

$$K^-(|k-l|) = \min_{\substack{i\in\omega_k \\ j\in\omega_l}} K(\gamma|i-j|). \tag{4.4.25}$$

Repeating the arguments leading to eqn (4.4.21) then results in the lower bound

$$Z_N(\beta, H, \gamma) \geqslant \exp[N f_N^-(m)] \tag{4.4.26}$$

where

$$f_N^-(m) = \beta[\alpha_L^- m^2 + Hm + \alpha_L^- - \alpha_N] + s^{-1} \ln A(m) \tag{4.4.27}$$

and

$$\alpha_L^- = (2L)^{-1}\gamma s \sum_{k,l=1}^{L} K^-(|k-l|). \tag{4.4.28}$$

In order to proceed with the thermodynamic limit we need the result (see Problem 5)

$$\lim_{M\to\infty} (2M)^{-1} \sum_{k\neq l=1}^{M} f(|k-l|) = \sum_{k=1}^{\infty} f(k) \tag{4.4.29}$$

which holds for any function f for which the series on the right-hand side of eqn (4.4.29) converges. It follows immediately, from eqns (4.4.19) and (4.4.29), that

$$\lim_{N\to\infty} \alpha_N = \gamma \sum_{l=0}^{\infty} K(\gamma l) \tag{4.4.30}$$

and also, from eqn (4.4.28), that for fixed s

$$\lim_{L\to\infty} \alpha_L^- = \alpha^- \tag{4.4.31}$$

where

$$\alpha^- = \gamma s \sum_{k=0}^{\infty} K^-(k). \tag{4.4.32}$$

Having taken the thermodynamic limit ($N, L \to \infty$ with $s = N/L$ fixed), we now note from the definitions of K^+ and K^- that α^+ and α^- are upper and lower Riemann sums and that

$$\lim_{\gamma\to 0+} \alpha^+ = \lim_{\gamma\to 0+} \alpha^- = \lim_{\gamma\to 0+} \lim_{N\to\infty} \alpha_N$$
$$= \lim_{\gamma\to 0+} \gamma \sum_{l=0}^{\infty} K(\gamma l)$$
$$= \int_0^\infty K(x)\mathrm{d}x \equiv \alpha/2. \tag{4.4.33}$$

Combining all of the above results, we then have

$$\max_{|m|\leqslant 1} f_s(m) \leqslant -\beta\psi(\beta, H; s) \leqslant s^{-1} \ln(1+2s) + \max_{|m|\leqslant 1} f_s(m) \tag{4.4.34}$$

where

$$-\beta\psi(\beta, H; s) = \lim_{\gamma\to 0+} \lim_{\substack{N,L\to\infty \\ s=N/L \text{ fixed}}} N^{-1} \ln Z_N(\beta, H, \gamma) \tag{4.4.35}$$

and

$$f_s(m) = \beta\left(\frac{\alpha m^2}{2} + Hm\right) + s^{-1} \ln A(m). \tag{4.4.36}$$

Since s is now arbitrary, we can take it to be arbitrarily large in which case the upper and lower bounds in eqn (4.4.34) coalesce to give the result

$$-\beta\psi(\beta, H) = \max_{|m|\leqslant 1} f(m) \tag{4.4.37}$$

where

$$f(m) = \frac{\beta\alpha m^2}{2} + \beta Hm + \lim_{s\to\infty} s^{-1} \ln A(m) \tag{4.4.38}$$

and, on using Stirling's formula $M! \sim M^M \exp(-M)$ which is valid for large M, we have from the definition in eqn (4.4.20) of $A(m)$ that

$$\lim_{s\to\infty} s^{-1} \ln A(m) = -\tfrac{1}{2}(1+m)\ln \tfrac{1}{2}(1+m) - \tfrac{1}{2}(1-m)\ln \tfrac{1}{2}(1-m). \tag{4.4.39}$$

It is left as an exercise (Problem 6) to show that after the maximization in eqn (4.4.37) has been performed one recovers the required Curie–Weiss result eqn (4.4.6) with $K = \beta\alpha$.

4.5 Characterization of phase transitions by critical exponents

In Sections 4.2 and 4.3, we saw that the classical van der Waals and Curie–Weiss theories for gas–liquid and magnetic transitions, respectively, predict similar critical behaviour which we summarize as follows:

1. The specific heats at constant volume C_v and at zero field C_0 have jump discontinuities at $T = T_c$.

2. The coexistence curves are parabolic. That is,

$$v_g - v_\ell = \left(1 - \frac{T}{T_c}\right)^{1/2}, \quad m_0 \sim \left(1 - \frac{T}{T_c}\right)^{1/2} \quad \text{as } T \to T_c - \tag{4.5.1}$$

where v denotes specific volume, m_0 spontaneous magnetization, and the subscripts g and ℓ refer to gas and liquid, respectively.

3. The isothermal compressibility K_T and the zero-field isothermal susceptibility χ_0 have simple poles at $T = T_c$. That is,

$$K_T \sim \left|1 - \frac{T}{T_c}\right|^{-1}, \quad \chi_0 \sim \left|1 - \frac{T}{T_c}\right|^{-1} \quad \text{as } T \to T_c. \tag{4.5.2}$$

4. The critical isotherms are cubic. That is,

$$p - p_c \sim (v_c - v)^3 \quad \text{at } T = T_c \text{ and } v \to v_c$$

and

$$H \sim m^3 \quad \text{at } T = T_c \text{ as } H \to 0 \tag{4.5.3}$$

where p denotes pressure, H external magnetic field, and m magnetization.

The exponents 1/2 in eqn (4.5.1), -1 in eqn (4.5.2), and 3 in eqn (4.5.3) are particular examples of critical exponents which we define, in general, as follows

The critical exponent λ corresponding to the critical point $x = 0+$ of a function f is defined by

$$\lambda = \lim_{x \to 0+} \frac{\ln |f(x)|}{\ln |x|} \tag{4.5.4}$$

and we write

$$f(x) \sim x^\lambda \quad \text{as } x \to 0+ \tag{4.5.5}$$

whenever the limit in eqn (4.5.4) *exists.*

In the event that the limit of the ratio in eqn (4.5.4) exists when $x \to 0-$ we will denote the 'left-hand' exponent by λ' and write

$$f(x) \sim (-x)^{\lambda'} \quad \text{as } x \to 0-. \tag{4.5.6}$$

Note that a function with a jump discontinuity (or in fact no singularity) at $x = 0$ has critical exponents $\lambda = \lambda' = 0$ as does the function $\ln |x|$ (and other such functions possessing weak singularities). In order to distinguish between these two cases in particular, we will usually use the notation 0_{disc} and $0_{\ln}$ for critical exponents corresponding to a jump discontinuity and logarithmic divergence, respectively.

The standard definitions of some critical exponents are given in Table 4.1 for a number of corresponding thermodynamic quantities for fluid and magnetic systems which are specified above in our summary of the classical

Table 4.1 Definition of some critical exponents for fluid and magnetic systems with $\tau = \dfrac{T}{T_c} - 1$

Exponent	Fluid	Magnet
α	$C_v \sim \tau^{-\alpha}$ as $\tau \to 0+$	$C_0 \sim \tau^{-\alpha}$ as $\tau \to 0+$
α'	$C_v \sim (-\tau)^{-\alpha'}$ as $\tau \to 0-$	$C_0 \sim (-\tau)^{-\alpha'}$ as $\tau \to 0-$
β	$v_g - v_\ell \sim (-\tau)^\beta$ as $\tau \to 0-$	$m_0 \sim (-\tau)^\beta$ as $\tau \to 0-$
γ	$K_T \sim \tau^{-\gamma}$ as $\tau \to 0+$	$\chi_0 \sim \tau^{-\gamma}$ as $\tau \to 0+$
γ'	$K_T \sim (-\tau)^{-\gamma'}$ as $\tau \to 0-$	$\chi_0 \sim (-\tau)^{-\gamma'}$ as $\tau \to 0-$
δ	$p - p_c \sim \operatorname{sgn}(v - v_c)\lvert v - v_c \rvert^\delta$ at $\tau = 0$ as $p \to p_c$	$H \sim \operatorname{sgn}(m)\lvert m \rvert^\delta$ at $\tau = 0$ as $H \to 0$

theories. A more complete list can be found in Fisher (1967) and Stanley (1971).

We note in passing that although the strict thermodynamic analogue of C_v is the specific heat at constant magnetization C_m, it is customary to define the exponents α and α' for magnetic systems in terms of the specific heat in zero field $C_{H=0}$ which we denote, for simplicity, by C_0. There is, of course, no reason to suppose that, in general, C_m and C_H have the same critical behaviour.

For the classical van der Waals and Curie–Weiss theories we have from the above that

$$\alpha = \alpha' = 0_{\text{disc}}, \quad \beta = 1/2, \quad \gamma = \gamma' = 1 \quad \text{and} \quad \delta = 3. \tag{4.5.7}$$

At first sight it is somewhat remarkable that the two classical theories have the same set of critical exponents. It is even more remarkable that there is close agreement between corresponding critical exponent values for real systems, both fluid and magnetic, as indicated in Table 4.2.

The experimental values listed in Table 4.2 actually represent a rather biased average of values obtained for various fluid and magnetic systems. Specific heat exponents, for example, are typically small and in the range 0_{ln} to 0.2 (and it should be noted that it is numerically difficult to distinguish a small power divergence from a logarithmic divergence). On the other hand, the gases Ar, Kr, Ne, Xe, N_2, O_2, CO, and CH_4, analysed by Guggenheim (1945) all have $\beta = 0.33 \pm 0.01$. Similar values are obtained for a wide variety of magnetic systems but there are some exceptions such as $CrBr_3$ which has $\beta \approx 0.37$. There is some variation in the other exponents and γ' for magnetic systems in particular is extremely difficult to estimate. More experimental

Table 4.2 Comparison of observed critical exponents with those of the classical theories

Exponent	Classical theories	Experimental (fluid and magnetic)
α	0_{disc}	0.15
α'	0_{disc}	0.15
β	1/2	0.33
γ	1	1.3
γ'	1	1.2
δ	3	4.5

detail on critical exponents can be found in Stanley (1971) or the review article by Heller (1967).

It will be noted from Table 4.2 that there is some discrepancy between experimental values of critical exponents and the values obtained from the classical theories. Away from the critical point the classical theories agree quite well with experiment but they obviously break down as one approaches the critical point. This can be understood in the light of the discussion in the previous section where we have shown that the classical theories are valid in the limit of infinitely weak long-range interactions. When the range of the potential (γ^{-1}) is finite, one might then expect deviations from classical behaviour in some temperature range of order γ around the critical point. In fact even for short-range potentials the effective range of interaction, as measured by the number of particles interacting with a given particle, can be relatively large. For example, on a face-centred cubic lattice each site has 12 nearest neighbours so that even with nearest-neighbour interaction the effective range of interaction is quite large.

Another feature reflected in Table 4.2, which is more interesting than the breakdown of the classical theories, is the fact that critical exponents appear to fall into *universality classes* and to be rather insensitive to the detailed microscopic properties of the system in question. Some progress has been made in isolating the apparent small number of parameters that characterize universality classes or on which critical exponents depend (such as the dimensionality of the system and certain symmetries of the order parameter) but there are still many unsolved problems and questions in this area. We will have more to say on this point in later chapters.

When the study of critical exponents became fashionable in the 1960s people began to look for possible relations between them. The first such relation

$$\alpha' + 2\beta + \gamma' = 2 \tag{4.5.8}$$

was, in fact, conjectured by Essam and Fisher (1963) on the basis of a particular droplet model of condensation. Shortly afterwards Rushbrooke (1963) pointed out that from thermodynamics alone one has the inequality

$$\alpha' + 2\beta + \gamma' \geqslant 2. \tag{4.5.9}$$

The inequality

$$\alpha' + \beta(1+\delta) \geqslant 2 \tag{4.5.10}$$

obtained by Griffiths (1965*a*), was followed by a flood of new critical exponents, some measurable and others not, and various inequalities connecting them. The interested reader is referred to Stanley (1971) for details.

It will be noted from eqn (4.5.7) that, in fact, the inequalities (4.5.9) and (4.5.10) are satisfied as equalities for the classical critical exponents, and that from Table 4.2 the experimental values are also consistent with equality. We will see in subsequent chapters that various exact and approximate results support the conjectured equalities in eqns (4.5.9) and (4.5.10) which, as we will see in Chapter 7, are consequences of scaling and renormalization-group arguments.

With hindsight, the inequalities are not at all difficult to derive. Thus, if we start with the thermodynamic identity

$$C_H - C_M = T(\partial M/\partial T)_H^2 (\partial M/\partial H)_T^{-1}, \tag{4.5.11}$$

use the fact that $C_M \geqslant 0$, and take the limit $H \to 0+$, we obtain, for $\tau = T/T_c - 1 < 0$,

$$\ln C_0 \geqslant 2\ln(\partial m_0/\partial T) - \ln \chi_0 + \ln T. \tag{4.5.12}$$

For small $|\tau|$ it follows that

$$\frac{\ln C_0}{\ln|\tau|} \leqslant \frac{2\ln(\partial m_0/\partial T)}{\ln|\tau|} - \frac{\ln \chi_0}{\ln|\tau|} + \frac{\ln T}{\ln|\tau|} \tag{4.5.13}$$

and hence, from eqn (4.5.4) and the definitions given in Table 4.1, we have on taking the limit $\tau \to 0-$ the inequality

$$-\alpha' \leqslant 2(\beta - 1) + \gamma', \tag{4.5.14}$$

which is only a slightly disguised form of eqn (4.5.9).

The Griffiths inequality (4.5.10) can also be obtained from thermodynamic arguments (Rushbrooke 1965) or, alternatively, by convexity arguments (Griffiths 1965*a*; see Problem 9).

4.6 The Yang–Lee theory of phase transitions

In two classic papers published in 1952 (Yang and Lee 1952; Lee and Yang 1952), Yang and Lee presented a theory of phase transitions based on the properties and distribution of zeros of the grand-canonical partition function in the complex z or fugacity plane.

Recall that the grand-canonical partition function Z_G, discussed in Section 2.7, is the generating function for the canonical partition function. That is,

$$Z_G(V, T, z) = \sum_{N=0}^{\infty} z^N Z(V, N, T) \tag{4.6.1}$$

where

$$Z(V, N, T) = (N!)^{-1} \int_V \cdots \int \exp[-\beta\Phi(\mathbf{r}_1, \mathbf{r}_2, \ldots, \mathbf{r}_N)] \mathrm{d}\mathbf{r}_1 \cdots \mathrm{d}\mathbf{r}_N \tag{4.6.2}$$

is the canonical partition function for a system of N particles contained in a region with volume V and with interaction potential Φ. The variable z is called the activity or fugacity.

The grand-canonical potential is defined (in the thermodynamic limit) by

$$\chi(T, z) = \lim_{V \to \infty} V^{-1} \ln Z_G(V, T, z) \tag{4.6.3}$$

and, in terms of χ, the pressure p and specific volume v are given, respectively, by

$$p = kT\chi(T, z) \tag{4.6.4}$$

and

$$v = \left(z \frac{\partial \chi}{\partial z}\right)^{-1}. \tag{4.6.5}$$

The equation of state is obtained by eliminating z between eqns (4.6.4) and (4.6.5).

Yang and Lee considered a gas of hard-sphere particles and showed that under certain conditions on the interaction potential (as discussed in Section 3.7) the limit in eqn (4.6.3) exists, is continuous and monotonically increasing for real z, and that quantities obtained by differentiating χ are well-defined in certain regions of the complex z-plane.

The simplifying feature of the hard-sphere condition, as noted previously, is that

$$Z(V, N, T) = 0 \qquad \text{when } N > M(V) \tag{4.6.6}$$

where $M(V)$ is the maximum number of hard spheres that can be packed into the given region of volume V. This result implies that the grand-canonical partition function (4.6.1) is a polynomial in z of degree $M(V)$. Adopting the convention $Z(V, 0, T) = 1$, we can then write

$$Z_G(V, T, z) = \prod_{r=1}^{M(V)} \left(1 - \frac{z}{z_r}\right) \tag{4.6.7}$$

where $z_1, z_2, \ldots, z_{M(V)}$ are the zeros of $Z_G(V, T, z)$ in the complex z-plane. It is to be noted from eqn (4.6.2) that $Z(V, N, T) \geqslant 0$ and hence from eqn (4.6.1) that none of the zeros of Z_G can be real and positive. This is simply a reflection of the fact that for a finite system there is no phase transition. Yang and Lee proved that this is also the case for the infinite system provided no real positive z is an accumulation point of zeros of Z_G. In fact they proved much more, namely, that if any region R in the complex plane is free of zeros in the thermodynamic limit, then $\chi(z, T)$ is an analytic function of z for z in R.

In order to see how a phase transition develops in the Yang–Lee theory we first note that $M(V)$ diverges as $V \to \infty$. In the event that the zeros of Z_G coalesce in the limit $V \to \infty$ to form curves C in the complex plane, the grand-canonical potential from eqns (4.6.3) and (4.6.7) will have the form

$$\chi(z, T) = \lim_{V \to \infty} V^{-1} \sum_{r=1}^{M(V)} \ln\left(1 - \frac{z}{z_r}\right) \tag{4.6.8}$$

$$= \int_{\mathrm{C}} \ln\left(1 - \frac{z}{s}\right) g(s) \mathrm{d}s \tag{4.6.9}$$

where $Vg(s)$ is the (asymptotic) density of zeros, so that, in particular,

$$\lim_{V \to \infty} \frac{M(V)}{V} = \int_{\mathrm{C}} g(s) \mathrm{d}s. \tag{4.6.10}$$

One possible situation, illustrated in Fig. 4.9, is for the zeros (which must occur in complex conjugate pairs) to form the single curve C in the limit $V \to \infty$ which crosses the real axis at some point of accumulation $z_c > 0$. From Yang and Lee's first result, $\chi(z, T)$ and hence, from eqn (4.6.4), the pressure p

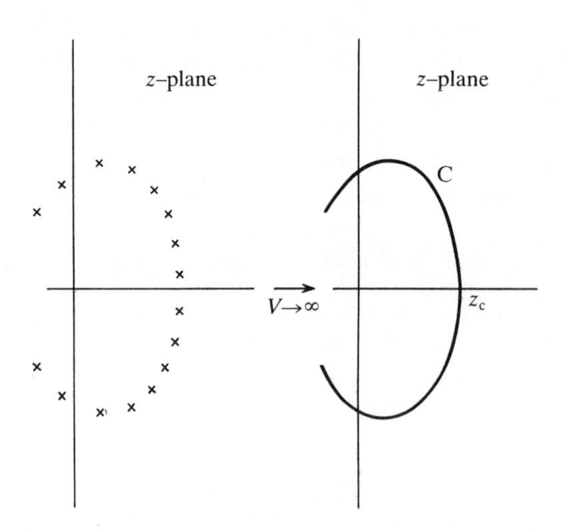

FIG. 4.9 Distribution of zeros of Z_G in the thermodynamic limit.

must be continuous and monotonic but may have a kink at z_c as shown in Fig. 4.10. In this case v^{-1} from eqn (4.6.5) will have a jump discontinuity at z_c. Graphical elimination of z in Fig. 4.10 then results in the isotherm in Fig. 4.11, corresponding to the usual gas–liquid condensation.

The above situation is, of course, not the only possibility. There could be more than one, or even an infinite number, of accumulation points of zeros on the positive real axis, so that, in principle, the Yang–Lee theory can account for any conceivable type of phase transition. In this sense it is more of a characterization of a phase transition than a theory since it doesn't really address itself to the fundamental question of obtaining conditions on the

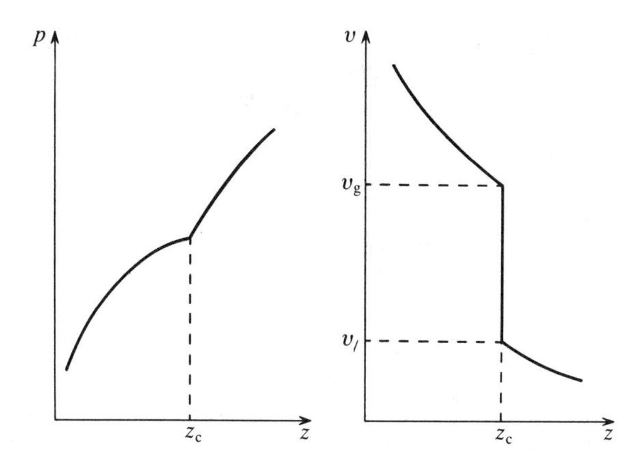

FIG. 4.10 Pressure p and specific volume v as functions of z.

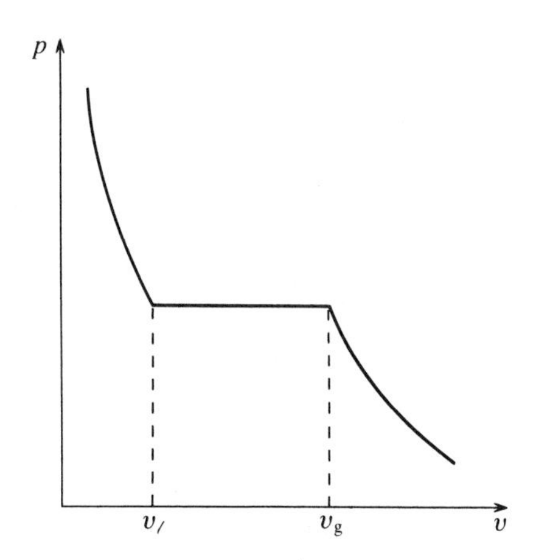

FIG. 4.11 Isotherm obtained by eliminating z graphically in Fig. 4.10.

underlying microscopic description, such as dimensionality, range of interaction, symmetries, and so forth, that will guarantee the existence of a phase transition.

The theory has, however, been successful in finding conditions under which there is no phase transition by showing that in certain circumstances, for example, at sufficiently high temperatures, the real axis is free from zeros in the thermodynamic limit. There is also the celebrated Lee–Yang circle theorem which states that the zeros of the grand-canonical partition function for the lattice gas, defined in Section 3.4, with attractive interactions only, lie on the unit circle in the complex fugacity plane. It follows that there can be only one transition point (corresponding to $z_c = 1$ in the above notation) and that the transition can be expected to be of the usual form for gas–liquid condensation. In magnetic language $z_c = 1$ corresponds to $H = 0$ so that the Lee–Yang circle theorem in this case states that an Ising model with attractive interactions can only have one transition and that it can only occur in zero field. The interested reader is referred to Ruelle (1969) or the original work of Yang and Lee.

4.7 Peierls's argument for the existence of a phase transition

In the previous section, we briefly indicated how Yang–Lee type arguments can be used to establish the non-existence of phase transitions in certain circumstances. In order to establish the existence of a phase transition one must, according to our definition given in Section 4.1, show that the free energy in the thermodynamic limit is singular, which can usually only be established by an explicit evaluation of the free energy. As we will see in Chapter 6 this has only been done in very special cases.

Alternative arguments to establish the existence of a phase transition date back to Peierls (1936*a*) who was the first to show that the spontaneous magnetization for the two-dimensional Ising model on a square lattice with nearest-neighbour interactions only was non-zero for sufficiently low temperatures. Strictly speaking, one also needs to show that the spontaneous magnetization is identically zero for sufficiently high temperatures to deduce the existence of a singular point or phase transition. This can be done in certain circumstances but we will not enter into any details here.

The original argument given by Peierls is in fact incomplete. A correct form of the argument was given by Griffiths (1964) and we will follow his later treatment (Griffiths 1972) here.

We consider a square lattice of N spins $\mu_i = \pm 1$, $i = 1, 2, \ldots, N$ and with interaction energy

$$E\{\mu\} = -J \sum_{(ij)} \mu_i \mu_j - H \sum \mu_i \tag{4.7.1}$$

in a given configuration $\{\mu\} = (\mu_1, \mu_2, \ldots, \mu_N)$ of spins. The sum over (ij)

denotes a sum over nearest-neighbour lattice sites i and j and we will assume that the system is ferromagnetic, i.e. $J>0$.

The partition function is defined by

$$Z_N(\beta, H)=\sum_{\{\mu\}} \exp(-\beta E\{\mu\}) \tag{4.7.2}$$

and the average magnetization per spin by

$$\begin{aligned} m_N(\beta, H) &= \left\langle N^{-1} \sum_{i=1}^{N} \mu_i \right\rangle \\ &= Z_N^{-1} \sum_{\{\mu\}} \left(N^{-1} \sum_{i=1}^{N} \mu_i \right) \exp(-\beta E\{\mu\}). \end{aligned} \tag{4.7.3}$$

The spontaneous magnetization m_0 is defined by

$$m_0(\beta)= \lim_{H\to 0+} \lim_{N\to\infty} m_N(\beta, H) \tag{4.7.4}$$

where, by symmetry, it is to be noted that the order of the limits in eqn (4.7.4) is important. That is, from eqns (4.7.1) and (4.7.2) and the fact that $E\{\mu\}=E\{-\mu\}$ when $H=0$, $m_N(\beta, 0)=0$ for all β.

Instead of examining $m_0(\beta)$ directly we consider $\hat{m}_N(\beta)$ defined by

$$\hat{m}_N(\beta, H)=\left\langle N^{-1} \sum_{i=1}^{N} \mu_i \right\rangle^{+} \tag{4.7.5}$$

where $\langle \cdot\cdot \rangle^{+}$ denotes the thermal average with 'plus boundary conditions'.

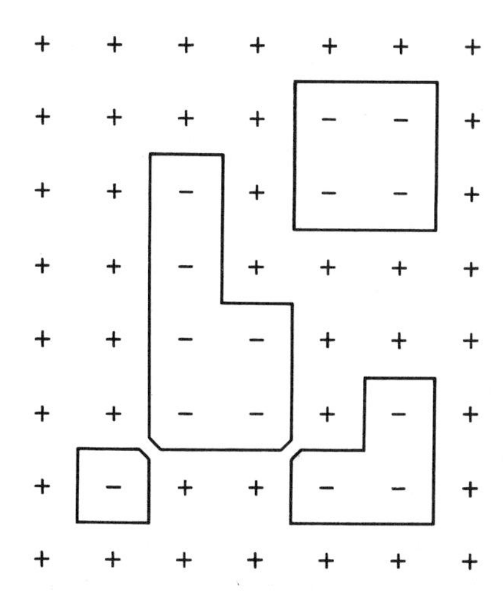

FIG. 4.12 Borders separating up (+) and down (−) spins on a lattice with '+' boundary conditions.

That is, the sum over configurations in Z_N (eqn (4.7.2)) and eqn (4.7.3) have boundary spins restricted to $\mu_i = +1$, as shown in Fig. 4.12. The symmetry argument above no longer applies in this case and for all finite N it is obvious that $\hat{m}_N(\beta, 0) \neq 0$.

We now proceed in two steps: first showing that there exists β_1 such that

$$\hat{m}_N(\beta, 0) \geqslant \alpha > 0 \qquad \text{for } \beta > \beta_1 \tag{4.7.6}$$

with α independent of N. We then show from convexity arguments that the spontaneous magnetization in eqn (4.7.4) is also bounded below by α for $\beta > \beta_1$.

For a given configuration with plus boundary conditions, also shown in Fig. 4.12, we construct closed polygons or borders by drawing lines between pairs of unlike spins. When two or more such polygons meet at a corner we separate them by cutting the corners adjacent to the sites with down spins, as shown in Fig. 4.12. A polygon with perimeter b consists of b line segments and obviously cannot enclose more than $(b/4)^2$ lattice sites.

We now let $n(b)$ be the number of ways of drawing a polygon of perimeter b on the lattice and define the characteristic function $\chi_b^{(i)}\{\mu\}$ by

$$\chi_b^{(i)}\{\mu\} = \begin{cases} 1 & \text{if the } i\text{th polygon with perimeter } b \text{ is present in } \{\mu\} \\ 0 & \text{otherwise.} \end{cases} \tag{4.7.7}$$

The number of down spins ($\mu = -1$) in configuration $\{\mu\}$ with plus boundary conditions is then bounded by

$$N_-\{\mu\} \leqslant \sum_{b=4,6,\ldots} (b/4)^2 \sum_{i=1}^{n(b)} \chi_b^{(i)}\{\mu\}. \tag{4.7.8}$$

Since

$$N^{-1} \sum_{i=1}^{N} \mu_i = 1 - 2N_-\{\mu\}/N, \tag{4.7.9}$$

it then follows, from eqn (4.7.5), that

$$\hat{m}_N(\beta, 0) \geqslant 1 - \frac{2}{N} \sum_{b=4,6,\ldots} (b/4)^2 \sum_{i=1}^{n/b)} \langle \chi_b^{(i)}\{\mu\} \rangle^+. \tag{4.7.10}$$

Now, by definition,

$$\langle \chi_b^{(i)}\{\mu\} \rangle^+ = \sum_{\{\mu\}}{}^{+\prime} \exp(-\beta E\{\mu\}) \Big/ \sum_{\{\mu\}}{}^{+} \exp(-\beta E\{\mu\}) \tag{4.7.11}$$

where the plus superscript denotes a sum over configurations with plus boundary conditions and the prime denotes a sum over all configurations in which the specified polygon appears. We associate with each configuration $\{\mu\}$ in which the polygon occurs a configuration $\{\mu\}^*$ obtained by replacing

μ_i by $-\mu_i$ for all spins inside the polygon. In zero field the energies of the two configurations, from eqn (4.7.1), differ by

$$E\{\mu\} - E\{\mu\}^* = 2bJ. \tag{4.7.12}$$

An upper bound to eqn (4.7.11) is then obtained by restricting the configurations in the denominator to those of type $\{\mu\}^*$ yielding

$$\langle \chi_b^{(i)}\{\mu\}\rangle^+ \leqslant \exp(-2\beta bJ). \tag{4.7.13}$$

To obtain an upper bound on $n(b)$ we first note that there are at most N polygons of the same type, where we say two polygons are of the same type if one can be obtained by a rigid translation on the lattice. A polygon of a given type having perimeter b may be constructed by attaching b lines end-to-end on a square lattice. Since the position of the first line is arbitrary and there are at most three possibilities for the remaining $b-1$ lines we have the bound

$$n(b) \leqslant N\,3^{b-1}. \tag{4.7.14}$$

Combining inequalities (4.7.10), (4.7.13), and (4.7.14) and allowing b to range over all positive integers results in the bound

$$\begin{aligned} \hat{m}_N(\beta, 0) &\geqslant 1 - (1/24) \sum_{b=0}^{\infty} b^2 [3 \exp(-2\beta J)]^b \\ &= 1 - r(1+r)/24(1-r)^3 \end{aligned} \tag{4.7.15}$$

provided

$$r = 3 \exp(-2\beta J) < 1. \tag{4.7.16}$$

Furthermore, since r can be made arbitrarily small by choosing β sufficiently large, the required result, eqn (4.7.6), follows immediately from eqn (4.7.15).

In order to show that the spontaneous magnetization (4.7.4) is also non-zero at sufficiently low temperatures we note first of all that, since the free energy for a spin system is, in general, a concave function of H, we have that, for $H > 0$,

$$[\hat{\psi}_N(\beta, H) - \hat{\psi}_N(\beta, 0)]H^{-1} \leqslant \left.\frac{\partial \hat{\psi}_N}{\partial H}\right|_{H=0} = -\hat{m}_N(\beta, 0) \tag{4.7.17}$$

where $\hat{\psi}_N(\beta, H)$ denotes the free energy per spin of the finite lattice with plus boundary conditions.

Since the boundary conditions contribute only a 'surface term' to the free energy,

$$\lim_{N \to \infty} \hat{\psi}_N(\beta, H) = \psi(\beta, H) \tag{4.7.18}$$

where $\psi(\beta, H)$ is the limiting free energy of the system without boundary conditions. Finally, in the limit $H \to 0+$, the left-hand side of eqn (4.7.17) approaches $-m_0(\beta)$ and the required result follows from eqn (4.7.6).

4.8 Problems

1. Show that the van der Waals gas has a critical point when

$$p_c = \alpha/54v_0^2, \qquad v_c = 3v_0, \qquad T_c = 4\alpha/27v_0k$$

and that, in terms of the scaled variables,

$$\tilde{p} = p/p_c, \qquad \tilde{v} = v/v_c, \qquad \tilde{T} = T/T_c,$$

the van der Waals equation of state has the universal form

$$\tilde{p} = 8\tilde{T}(3\tilde{v} - 1)^{-1} - 3\tilde{v}^{-2}.$$

2. Show that C_v on the critical isochore of the van der Waals gas has a jump discontinuity at $T = T_c$ of magnitude $9k/2$.

3. Show, from eqn (4.3.30), that the zero-field specific heat for the Curie–Weiss model has a jump discontinuity at $T = T_c$ of magnitude $3k/2$.

4. Show that the specific heat at constant magnetization for the Curie–Weiss model is identically zero.

5. Prove that

$$\lim_{M \to \infty} (2M)^{-1} \sum_{k \neq l = 1}^{M} f(|k - l|) = \sum_{k=1}^{\infty} f(k)$$

holds for any function f for which the series on the right-hand side of the equation converges.

6. Show that

$$\max_{|x| \leq 1} \{\tfrac{1}{2}Kx^2 + \beta Hx - \tfrac{1}{2}(1+x)\ln\tfrac{1}{2}(1+x) - \tfrac{1}{2}(1-x)\ln\tfrac{1}{2}(1-x)\}$$

$$= -\tfrac{1}{2}Km^2 + \ln 2\cosh(Km + \beta H)$$

where m is a solution of the mean-field equation

$$m = \tanh(Km + \beta H).$$

7. From the definition of the critical exponent λ of a function f given in eqns (4.5.4) and (4.5.5), show that:

(a) if $f(x) \sim x^{\lambda}$ and $g(x) \sim x^{\mu}$ as $x \to 0+$ and $f(x) < g(x)$ for sufficiently small positive x, then $\lambda > \mu$;

(b) if $f(x) \sim x^{\lambda}$ as $x \to 0+$ and $\lambda > -1$, then

$$\int_0^x f(t)\mathrm{d}t \sim x^{\lambda+1} \qquad \text{as } x \to 0+;$$

(c) if $f(x) \sim x^{\lambda}$ and $g(x) \sim x^{\mu}$ as $x \to 0+$, then $f(g(x)) \sim x^{\lambda\mu}$ as $x \to 0+$.

8. Show, from the thermodynamic relation given in eqn (4.5.11), that $\alpha' + 2\beta + \gamma' = 2$ if $\lim_{T \to T_c-} C_M/C_H \neq 1$.

9. Using the fact that the free energy $\psi(m, T)$ considered as a function of magnetization m and temperature T is a concave function of T for fixed m show that

$$\psi(m_0, T_c) - \psi(0, T_c) \leq (T_c - T)[s(0, T_c) - s(0, T)]$$

where m_0 is the spontaneous magnetization at $T<T_c$ and $s=-(\partial\psi/\partial T)_m$ is the entropy.

From the definition of critical exponents and properties given in Problem 7, deduce the Griffiths exponent inequality (4.5.10).

10. Consider the model partition function for integral N and V defined by

$$Z(V, N)=\sum_{k=0}^{V} V!/(V-N+k)!(N-k)!.$$

Show that:

(a) the grand-canonical partition function is given by

$$Z_G(V, z)=\sum_{N=0}^{\infty} z^N Z(V, N)=(z^{V+1}-1)(z+1)^V/(z-1);$$

(b) the grand-canonical potential $\chi(z)$ is given in the thermodynamic limit by

$$\chi(z)=\lim_{V\to\infty} V^{-1}\ln Z_G(V, z)=\begin{cases}\ln(z+1) & \text{when } |z|\leqslant 1\\ \ln z(z+1) & \text{when } |z|\geqslant 1;\end{cases}$$

(c) From the definitions of p and v, i.e.

$$\beta p=\chi(z) \qquad \text{and} \qquad v^{-1}=z\frac{\partial}{\partial z}\chi(z)$$

and the result of part (b), obtain p as a function of v. Deduce that

$$\beta p=\ln 2 \qquad \text{when } \tfrac{2}{3}\leqslant v\leqslant 2.$$

4.9 Solutions to problems

1. The van der Waals equation of state

$$\beta p=(v-v_0)^{-1}-\alpha/2kTv^2 \tag{1}$$

for $p=p_c$ and $T=T_c$ is equivalent to the cubic equation

$$v^3-\left(v_0+\frac{kT_c}{p_c}\right)v^2+\frac{\alpha}{2p_c}v-\frac{\alpha v_0}{2p_c}\equiv(v-v_c)^3=0.$$

Equating coefficients gives the stated results, and direct substitution of $p=p_c\tilde{p}$, $v=v_c\tilde{v}$, $T=T_c\tilde{T}$ in eqn (1) gives the universal form of the equation of state.

2. Define the scaled free energy $\tilde{\psi}(\tilde{v}, \tilde{T})$ by $\tilde{\psi}(\tilde{v}, \tilde{T})=(p_c v_c)^{-1}\psi(v, T)$ so that $\tilde{p}=-(\partial\tilde{\psi}/\partial\tilde{v})$ etc., and note that on the critical isochore, $\tilde{v}=1$,

$$\begin{aligned}\tilde{\psi}(1, \tilde{T})&\sim\tfrac{1}{2}[\psi_w(1+\varepsilon, \tilde{T})+\psi_w(1-\varepsilon, \tilde{T})]\\ &=\tilde{\psi}(1, 1)-\frac{\varepsilon^2}{2!}\frac{\partial\tilde{p}}{\partial\tilde{v}}-\frac{\varepsilon^4}{4!}\frac{\partial^3\tilde{p}}{\partial\tilde{v}^3}+\ldots \qquad \text{as } \tau=1-\tilde{T}\to 0+\end{aligned}$$

where $\varepsilon=2\tau^{1/2}$ and ψ_w denotes the van der Waals free energy, i.e. the right-hand side of eqn (4.2.17) without the convex envelop construction.

From the universal form of the equation of state,

$$\frac{\partial \tilde{p}}{\partial \tilde{v}} \sim 6\tau \quad \text{and} \quad \frac{\partial^2 \tilde{p}}{\partial \tilde{v}^3} \sim -9 \quad \text{at } \tilde{v}=1 \text{ as } \tau \to 0+.$$

The required result follows by noting that the jump in the specific heat at the critical point is given by

$$\Delta C_v = -\frac{p_c v_c}{T_c} \frac{\partial^2}{\partial \tilde{T}^2} \tilde{\psi}(1, \tilde{T})\bigg|_{\tilde{T}=1}.$$

3. From eqn (4.3.30) the jump in the zero-field specific heat at T_c is given by

$$\Delta C_0 = -\frac{J}{2} \frac{dm_0^2}{dT}\bigg|_{T=T_c} \tag{1}$$

where

$$m_0 = \tanh(Km_0) > 0 \qquad \text{for } K = \frac{J}{kT} > \frac{J}{kT_c} = 1.$$

From eqn (4.3.22)

$$m_0^2 \sim 3\left(1 - \frac{T}{T_c}\right) \qquad \text{as } T \to T_c. \tag{2}$$

Combining eqns (1) and (2) gives the required result.

4. The equation of state for the Curie–Weiss model

$$m = \tanh(\beta Jm + \beta H)$$

can be written in the form

$$H = kT \operatorname{artanh} m - Jm$$

$$= \frac{kT}{2} \ln\left(\frac{1+m}{1-m}\right) - Jm. \tag{1}$$

The free energy as a function of m and T can then be written as

$$\psi(T, m) = \tfrac{1}{2}Jm^2 - kT \ln 2\cosh(\beta Jm + \beta H)$$

with H given in terms of m and T by eqn (1).

It is an elementary matter to show that

$$\left(\frac{\partial \psi}{\partial T}\right)_m = -k \ln 2\cosh(\beta Jm + \beta H) + \frac{m}{T}(Jm + H) - m\left(\frac{\partial H}{\partial T}\right)_m$$

and then, from eqn (1), that

$$\left(\frac{\partial^2 \psi}{\partial T^2}\right)_m = -m\left(\frac{\partial^2 H}{\partial T^2}\right)_m \equiv 0.$$

The required result follows by noting that

$$C_m = -T\left(\frac{\partial^2 \psi}{\partial T^2}\right)_m.$$

It is an interesting exercise to check that the results

$$C_H = -(Jm+H)\left(\frac{\partial m}{\partial T}\right)_H$$

together with $C_m \equiv 0$ for the Curie–Weiss model, are consistent with the thermodynamic identity

$$C_m = C_H - T\left(\frac{\partial m}{\partial T}\right)_H^2 \Big/ \left(\frac{\partial m}{\partial H}\right)_T.$$

5.
$$S_M \equiv (2M)^{-1} \sum_{k \neq l=1}^{M} f(|k-l|) = M^{-1} \sum_{1 \leqslant k < l \leqslant M} f(l-k)$$

$$= M^{-1} \sum_{l=1}^{M-1} a_l$$

where

$$a_l = \sum_{k=1}^{l} f(k).$$

Let

$$a = \lim_{l \to \infty} a_l = \sum_{k=1}^{\infty} f(k).$$

Then, given $\varepsilon > 0$, there exists $N(\varepsilon)$ such that

$$a - \varepsilon < a_l < a + \varepsilon \qquad \text{for all } l > N(\varepsilon).$$

For M larger than $N(\varepsilon)$ we then have

$$M^{-1}\{A + [M-1-N(\varepsilon)](a-\varepsilon)\} < S_M < M^{-1}\{A + [M-1-N(\varepsilon)](a+\varepsilon)\}$$

where

$$A = \sum_{l=1}^{N(\varepsilon)} a_l.$$

Taking the limit $M \to \infty$ then gives

$$a - \varepsilon < \lim_{M \to \infty} S_M < a + \varepsilon$$

and since ε can be arbitrarily small the result follows.

6. The equation determining the maximum is

$$Kx + \beta H = \frac{1}{2}\ln\left(\frac{1+x}{1-x}\right) = \operatorname{artanh} x$$

which is equivalent to the mean-field equation

$$x \equiv m = \tanh(Km + \beta H).$$

The maximum value is then given by

$$\frac{1}{2}Km^2+\beta Hm-\frac{m}{2}\ln\left(\frac{1+m}{1-m}\right)-\frac{1}{2}\ln\frac{1}{4}(1-m^2)$$

$$=-\frac{1}{2}Km^2-\frac{1}{2}\ln\frac{1}{4}[1-\tanh^2(Km+\beta H)]$$

$$=-\frac{1}{2}Km^2+\ln 2\cosh(Km+\beta H).$$

7. By definition, $f(x)\sim x^{\lambda}$ as $x\to 0+$ if

$$\lambda=\lim_{x\to 0+}\frac{\ln f(x)}{\ln x}.$$

(a) Since $\ln x$ is negative for $0<x<1$ and $f(x)<g(x)$ for sufficiently small $x>0$,

$$\frac{\ln f(x)}{\ln x}>\frac{\ln g(x)}{\ln x}\qquad\text{as } x\to 0+.$$

Taking limits gives $\lambda>\mu$.
(b) Since

$$\int_0^x f(t)\mathrm{d}t\sim xf(x)\qquad\text{as } x\to 0+$$

$$\lim_{x\to 0+}\frac{\ln\left[\int_0^x f(t)\mathrm{d}t\right]}{\ln x}=\lim_{x\to 0+}\frac{\ln xf(x)}{\ln x}=1+\lambda.$$

(c) $$\lim_{x\to 0+}\frac{\ln f(g(x))}{\ln x}=\lim_{x\to 0+}\frac{\ln f(g(x))}{\ln g(x)}\cdot\frac{\ln g(x)}{\ln x}=\lambda\mu.$$

8. Dividing both sides of the thermodynamic relation

$$C_H-C_M=T\left(\frac{\partial M}{\partial T}\right)^2\Big/\left(\frac{\partial M}{\partial H}\right)>0$$

through by C_H and taking logarithms of both sides, we obtain in the limit $H\to 0+$ with $T<T_c$ the identity

$$\ln(1-C_{m_0}/C_0)=\ln T+2\ln|\partial m_0/\partial T|-\ln\chi_0-\ln C_0$$

where m_0 is the spontaneous magnetization, χ_0 is the zero-field isothermal susceptibility, and C_0 is the zero-field specific heat.

Dividing through by $\ln|T-T_c|$ and assuming $\lim_{T\to T_c-} C_{m_0}/C_0<1$, we then obtain, in the limit $T\to T_c-$, the identity

$$0=2(\beta-1)+\gamma'+\alpha',$$

which is the required result.

9. If we consider the free energy $\psi(m, T)$ to be a function of magnetization and temperature, the magnetic field and entropy are given by

$$H(m, T) = \left(\frac{\partial \psi}{\partial m}\right)_T \quad \text{and} \quad s(m, T) = -\left(\frac{\partial \psi}{\partial T}\right)_m.$$

Since $H(m, T) = 0$ when m is less than or equal to the spontaneous magnetization $m_0(T)$ at temperature T, it follows that

$$\psi(m, T) = \psi(0, T) \qquad \text{when } m \leqslant m_0(T).$$

Now, since $\psi(m, T)$ is a concave function of T for fixed m, it follows that for $T_1 < T_2 \leqslant T_c$ and arbitrary m,

$$-s(m, T_1) = \left.\frac{\partial \psi}{\partial T}\right|_{T_1} \geqslant \frac{\psi(m, T_2) - \psi(m, T_1)}{T_2 - T_1} \tag{1}$$

and

$$-s(m, T_2) = \left.\frac{\partial \psi}{\partial T}\right|_{T_2} \leqslant \frac{\psi(m, T_2) - \psi(m, T_1)}{T_2 - T_1}. \tag{2}$$

Setting $T_2 = T_c$, $T_1 = T$, and $m = m_0(T)$ in eqn (1) gives

$$\begin{aligned}\psi(m_0(T), T_c) &\leqslant \psi(m_0(T), T) + (T - T_c)s(m_0(T), T)\\ &= \psi(0, T) + (T - T_c)s(0, T).\end{aligned}$$

Similarly, if we set $T_2 = T_c$, $T_1 = T$ and $m = 0$ in eqn (2), we obtain

$$\psi(0, T) \leqslant \psi(0, T_c) + (T_c - T)s(0, T_c).$$

Combining the last two inequalities gives the required result

$$\psi(m_0(T), T_c) - \psi(0, T_c) \leqslant (T_c - T)[s(0, T_c) - s(0, T)]. \tag{3}$$

Now, when $T \to T_c-$, the left-hand side of eqn (3) by definition behaves as

$$[m_0(T)]^{1+\delta} \sim (T_c - T)^{\beta(1+\delta)}$$

and, since

$$C_0 = \left(\frac{\partial s}{\partial T}\right)_{H=0} \sim (T_c - T)^{-\alpha'} \qquad \text{as } T \to T_c-,$$

the right-hand side of eqn (3) behaves as

$$(T_c - T)(T_c - T)^{1-\alpha'}.$$

We therefore have a contradiction unless

$$\beta(1+\delta) \geqslant 2 - \alpha'$$

which is Griffiths's inequality (4.5.10).

10. (a)

$$\begin{aligned}(z^{V+1} - 1)(z+1)^V/(z-1) &= \sum_{m=0}^{V} z^m \sum_{l=0}^{V} \binom{V}{l} z^l\\ &= \sum_{N=0}^{2V} Z(V, N) z^N\end{aligned}$$

where

$$Z(V,N)=\begin{cases}\sum_{k=0}^{N} V!/(V-N+k)!(N-k)! & 0\leqslant N\leqslant V\\ \sum_{k=N-V}^{V} V!/(V-N+k)!(N-k)! & V+1\leqslant N\leqslant 2N.\end{cases}$$

The required result follows easily by noting that $(n!)^{-1}=0$ when n is a negative integer.

(b) The required result follows from part (a) by noting that as $V\to\infty$

$$Z_{\mathrm{G}}(V,z)\sim\begin{cases}(z+1)^{V} & \text{when } |z|\leqslant 1\\ z^{V}(z+1)^{V} & \text{when } |z|>1.\end{cases}$$

(c) From the result of part (b)

$$v^{-1}=\begin{cases}z(z+1)^{-1} & \text{when } |z|\leqslant 1\\ 1+(z+1)^{-1} & \text{when } |z|>1.\end{cases}$$

The required result then follows by noting that v undergoes a jump discontinuity from $v=\frac{2}{3}$ to $v=2$ at $z=1$.

5

FLUCTUATIONS AND CORRELATIONS

5.1 Introduction

In Chapter 2, we discussed the various Gibbs distributions or ensembles and the equivalence of their thermodynamic descriptions of large systems. In particular, we saw in Section 2.8 that, for large systems, the energy in the canonical description and the density in the grand-canonical description are approximately equal to their respective mean or average values. The root-mean-square deviations or *fluctuations*, of such quantities were shown to be small and of the order of the inverse square root of the size of the system. More precisely, we showed that, in the canonical description (2.8.8)

$$\left\langle \left(\frac{\mathscr{H}}{\langle \mathscr{H} \rangle} - 1 \right)^2 \right\rangle = N^{-1}(kT^2 C_V/E) \tag{5.1.1}$$

where $\mathscr{H}$ is the Hamiltonian of the system, $E = \langle \mathscr{H} \rangle / N$ is the average energy per particle, and C_V is the constant-volume specific heat per particle, and in the grand-canonical description (2.8.6),

$$\left\langle \left(\frac{N}{\langle N \rangle} - 1 \right)^2 \right\rangle = V^{-1}(kTK_T) \tag{5.1.2}$$

where K_T is the isothermal compressibility per particle. The left-hand sides of eqns (5.1.1) and (5.1.2) represent the mean-square deviation of energy and density and the angular brackets denote averages with respect to the appropriate Gibbs distribution.

In general, fluctuations in thermodynamic quantities such as energy and density are typically small for large systems, except, it will be noted from eqns (5.1.1) and (5.1.2), in the neighbourhood of a critical point where quantities such as C_V and K_T diverge. In the case of gas–liquid condensation, for example, the divergence of the compressibility corresponds to large density fluctuations which is observed as critical opalescence as one approaches the critical point.

To obtain some understanding of the microscopic behaviour of a system in the vicinity of a phase-transition point it is natural to study the way particles become correlated over large distances as one approaches a critical point. Viewed in this way, divergent thermodynamic quantities associated with a phase transition are a reflection of large fluctuations, in quantities such as the density which, in turn, are induced by long-range order or correlations between the particles.

In this chapter, we study fluctuations and correlations in the neighbourhood of critical points for fluid and magnetic systems.

5.2 Distribution functions and the fluctuation relation for fluids

In the Gibbs grand-canonical description one can define

$$p(\mathbf{r}_1, \ldots, \mathbf{r}_N, N) = [N! Z_G(V, T, z)]^{-1} z^N \exp[-\beta\Phi(\mathbf{r}_1, \ldots, \mathbf{r}_N)] \tag{5.2.1}$$

to be the probability density for a system contained in a region with volume V to have N particles with positions $\mathbf{r}_1, \ldots, \mathbf{r}_N$, where Φ is the interaction potential, z the fugacity, and

$$Z_G(V, T, z) = \sum_{N=0}^{\infty} z^N (N!)^{-1} \int \cdots \int_V \exp[-\beta\Phi(\mathbf{r}_1, \ldots, \mathbf{r}_N)] \mathrm{d}\mathbf{r}_1 \cdots \mathrm{d}\mathbf{r}_N \tag{5.2.2}$$

is the grand-canonical partition function.

The s-particle distribution function in the grand-canonical description is then defined in terms of p by

$$n_s(\mathbf{r}_1, \ldots, \mathbf{r}_s) = \sum_{N=s}^{\infty} [N!/(N-s)!] \int \cdots \int_V p(\mathbf{r}_1, \ldots, \mathbf{r}_N, N) \mathrm{d}\mathbf{r}_{s+1} \cdots \mathrm{d}\mathbf{r}_N \tag{5.2.3}$$

and may be interpreted as the probability density for any s particles to have positions $\mathbf{r}_1, \ldots, \mathbf{r}_s$ irrespective of the positions of the remaining particles.

It will be noted that, if one integrates the single-particle distribution function $n_1(\mathbf{r}_1)$ over $\mathbf{r}_1$, one immediately obtains, by definition,

$$\int_V n_1(\mathbf{r}_1) \mathrm{d}\mathbf{r}_1 = \langle N \rangle. \tag{5.2.4}$$

Similarly,

$$\iint_V n_2(\mathbf{r}_1, \mathbf{r}_2) \mathrm{d}\mathbf{r}_1 \mathrm{d}\mathbf{r}_2 = \langle N(N-1) \rangle \tag{5.2.5}$$

and so forth.

For simplicity, we will consider only uniform systems with periodic boundary conditions for which $n_1(\mathbf{r}_1)$ is independent of $\mathbf{r}_1$, $n_2(\mathbf{r}_1, \mathbf{r}_2)$ depends only on $\mathbf{r}_1 - \mathbf{r}_2$, etc. In such a case we have, from eqn (5.2.4), that

$$n_1(\mathbf{r}) = \langle N \rangle / V = \rho \qquad \text{for all } \mathbf{r} \tag{5.2.6}$$

is just the (mean) density, and that the *pair-correlation function*, defined in general by

$$G(\mathbf{r}_1, \mathbf{r}_2) = n_2(\mathbf{r}_1, \mathbf{r}_2) - n_1(\mathbf{r}_1)n_1(\mathbf{r}_2), \tag{5.2.7}$$

is a function only of $\mathbf{r}_1 - \mathbf{r}_2$.

Combining eqns (5.2.4), (5.2.5), and (5.2.7), we see that

$$\begin{aligned}\iint_V G(\mathbf{r}_1, \mathbf{r}_2)\,d\mathbf{r}_1\,d\mathbf{r}_2 &= \langle N^2\rangle - \langle N\rangle - \langle N\rangle^2 \\ &= \langle (N - \langle N\rangle)^2\rangle - \langle N\rangle \\ &= \langle N\rangle(\rho kTK_T - 1)\end{aligned} \tag{5.2.8}$$

where in the last step we have used eqn (5.1.2) relating the compressibility and the mean-square density fluctuation.

In the case of a uniform system with periodic boundary conditions, the left-hand side of eqn (5.2.8) becomes

$$\iint_V G(\mathbf{r}_1 - \mathbf{r}_2)\,d\mathbf{r}_1\,d\mathbf{r}_2 = V\int_V G(\mathbf{r})\,d\mathbf{r} \tag{5.2.9}$$

from which one obtains in the thermodynamic limit, $V \to \infty$ with $\rho = \langle N\rangle/V$ fixed, the so-called *fluctuation relation*

$$\rho kTK_T = 1 + \rho^{-1} \int G(\mathbf{r})\,d\mathbf{r} \tag{5.2.10}$$

which expresses the isothermal compressibility in terms of the integral (over all space) of the pair-correlation function.

Since K_T is finite away from a critical point it follows from eqn (5.2.10) that the integral of $G(\mathbf{r})$ must be finite and hence, from eqn (5.2.7), that

$$\lim_{|\mathbf{r}_1 - \mathbf{r}_2| \to \infty} G(\mathbf{r}_1 - \mathbf{r}_2) = \lim_{|\mathbf{r}_1 - \mathbf{r}_2| \to \infty} n_2(\mathbf{r}_1 - \mathbf{r}_2) - \rho^2 = 0. \tag{5.2.11}$$

That is,

$$n_2(\mathbf{r}_1 - \mathbf{r}_2) \sim n_1(\mathbf{r}_1)n_2(\mathbf{r}_2) \qquad \text{as } \mathbf{r}_1 - \mathbf{r}_2 \to \infty \tag{5.2.12}$$

and hence, for large separation, the particles become uncorrelated.

At a critical point, where K_T diverges, the integral of $G(\mathbf{r})$ must also diverge, which means that $G(\mathbf{r})$ becomes long-ranged in the sense that in d dimensions $G(\mathbf{r})$ decays more slowly than r^{-d} as $r \to \infty$. The precise nature of the decay of $G(\mathbf{r})$ at a critical point is of some interest and will be discussed more fully in subsequent sections.

The phenomenon of critical opalescence can also be understood from the fluctuation relation by noting that the Fourier transform of the pair-correlation function

$$\hat{G}(\mathbf{q}) = \int e^{i\mathbf{q}\cdot\mathbf{r}} G(\mathbf{r})\,d\mathbf{r} \tag{5.2.13}$$

is directly related to the scattering intensity $I(\mathbf{q})$ of the electromagnetic radiation with wave vector transfer $\mathbf{q}$ through the equation

$$\frac{I(\mathbf{q})}{I_0(\mathbf{q})}=1+\rho^{-1}\hat{G}(\mathbf{q}) \tag{5.2.14}$$

where $I_0(\mathbf{q})$ is the scattering intensity in the absence of correlations.

By comparing eqns (5.2.10), (5.2.13), and (5.2.14), we see that

$$\rho kTK_T=1+\rho^{-1}\hat{G}(\mathbf{0}) \tag{5.2.15}$$

which shows that the isothermal compressibility is proportional to the scattering intensity extrapolated to zero angle. It follows that as a critical point is approached, the amplitude of low-angle scattering becomes very large. The resulting large long-wavelength density fluctuations then produce the visible diffraction effect from the scattering of light known as critical opalescence.

5.3 The Ornstein–Zernike theory

The first theory of critical scattering was the classical theory of Ornstein and Zernike (1914). Like the theories of van der Waals and of Curie and Weiss discussed in Chapter 4, the Ornstein–Zernike theory provides a good description of reality away from a critical point and, in a certain technical sense, it is valid in the Kac limit of infinitely weak long-range potentials (Gates 1970). We will not enter into any rigorous discussion here on the validity of the Ornstein–Zernike theory. Instead, we will follow the heuristic development of the theory given by Fisher (1964*b*).

The basic idea in the Ornstein–Zernike theory is to think of correlations between two particles as being mediated by direct interactions between the two particles and indirect interactions via a third particle. We express this idea mathematically by writing the pair-correlation function in the form

$$G(\mathbf{r}_1-\mathbf{r}_2)=C(\mathbf{r}_1-\mathbf{r}_2)+\rho^{-1}\int C(\mathbf{r}_1-\mathbf{r}_3)G(\mathbf{r}_3-\mathbf{r}_2)\mathrm{d}\mathbf{r}_3. \tag{5.3.1}$$

The function $C(\mathbf{r})$ is called the *direct correlation function* and it is assumed that its range is of the same order as that of the interaction potential.

Equation (5.3.1) may be viewed simply as a definition of $C(\mathbf{r})$. Ornstein and Zernike, however, attached some physical significance to $C(\mathbf{r})$ and proposed calculating its mathematical form. We make no such attempt here and instead adopt the assumption that the form of $C(\mathbf{r})$ does not change drastically as one approaches a critical point.

If we take the Fourier transform of eqn (5.3.1), noting that the integral on the right-hand side of eqn (5.3.1) is a convolution of C and G we obtain

$$\hat{G}(\mathbf{q})=\hat{C}(\mathbf{q})+\rho^{-1}\hat{C}(\mathbf{q})\hat{G}(\mathbf{q}) \tag{5.3.2}$$

where $\hat{G}(\mathbf{q})$ is the Fourier transform of $G(\mathbf{r})$ defined by eqn (5.2.13) and

$$\hat{C}(\mathbf{q})=\int e^{i\mathbf{q}\cdot\mathbf{r}}C(\mathbf{r})\,d\mathbf{r}. \tag{5.3.3}$$

Solving eqn (5.3.2) for $\hat{G}(\mathbf{q})$ we obtain

$$1+\rho^{-1}\hat{G}(\mathbf{q})=[1-\rho^{-1}\hat{C}(\mathbf{q})]^{-1} \tag{5.3.4}$$

and, from the fluctuation relation written in the form (5.2.15), we have

$$1-\rho^{-1}\hat{C}(\mathbf{0})=(\rho kTK_T)^{-1}. \tag{5.3.5}$$

It follows that $\hat{C}(\mathbf{0})$ is defined for all T, including $T=T_c$ where K_T diverges and hence $\hat{C}(\mathbf{0})=\rho$ at the critical point.

We now make the crucial assumption, along with Ornstein and Zernike, that $C(\mathbf{r})$ is sufficiently short-ranged for all moments to exist and, moreover, that these moments exist right up to the critical point. Expanding the exponential in eqn (5.3.3) and using the symmetry of $C(\mathbf{r})$ in $\mathbf{r}$ we then obtain, for small $q=|\mathbf{q}|$.

$$\hat{C}(\mathbf{q})=\hat{C}(\mathbf{0})-R^2q^2+O(q^4) \tag{5.3.6}$$

where

$$R^2=\int[(\mathbf{q}/q)\cdot\mathbf{r}]^2C(\mathbf{r})\,d\mathbf{r}>0 \tag{5.3.7}$$

is proportional to the second moment of $C(\mathbf{r})$. It then follows from eqns (5.3.4) and (5.3.6) that

$$\begin{aligned}\hat{G}(\mathbf{q})&=\hat{C}(\mathbf{q})\,[1-\rho^{-1}\hat{C}(\mathbf{q})]^{-1}\\&=R^{-2}\hat{C}(\mathbf{0})\,[\kappa_1^2+q^2]^{-1}+O(q^2)\end{aligned} \tag{5.3.8}$$

where, from eqn (5.3.5),

$$\begin{aligned}\kappa_1^2&=R^{-2}(1-\rho^{-1}\hat{C}(\mathbf{0}))\\&=[R^2\rho kTK_T]^{-1}\geqslant 0.\end{aligned} \tag{5.3.9}$$

The simple Lorentzian form of $\hat{G}(\mathbf{q})$ in eqn (5.3.8) can be easily inverted to obtain the classical asymptotic Ornstein–Zernike result

$$G(r)\sim Ar^{-1}\exp(-\kappa_1 r) \qquad \text{as } r\to\infty \tag{5.3.10}$$

for three-dimensional systems.

It follows that correlations decay exponentially with inverse range κ_1, which, from eqn (5.3.9), goes to zero at the critical point. Assuming that R defined by eqn (5.3.7) is slowly varying near the critical point, and that K_T has the classical van der Waals simple pole singularity at T_c it follows from eqn (5.3.9) that the *correlation length*

$$\xi=\kappa_1^{-1}\sim C\left|1-\frac{T}{T_c}\right|^{-1/2} \qquad \text{as } T\to T_c \tag{5.3.11}$$

diverges at the critical point. Moreover, at the critical point, $T=T_c$, where

$\kappa_1 = 0$ the correlation function from eqn (5.3.10) has the simple power-law decay

$$G(\mathbf{r}) \sim A/r \qquad \text{at } T = T_c \qquad \text{as } r \to \infty. \tag{5.3.12}$$

The Fourier inversion of the Lorentzian form (5.3.8), which holds in any dimension d, can be easily inverted away from the critical point to obtain

$$G(\mathbf{r}) \sim A_d r^{-\frac{1}{2}(d-1)} \exp(-\kappa_1 r) \qquad \text{for fixed } \kappa_1 \text{ and } r \to \infty. \tag{5.3.13}$$

As noted by Fisher (1964*b*), a different asymptotic form results if we fix r (large) and let T approach T_c or, equivalently, $\kappa_1 \to 0+$. The results are (see Problem 3) for $d \geqslant 3$

$$G(\mathbf{r}) \sim B_c r^{-(d-2)} \exp(-\kappa_1 r) \qquad \text{for fixed } r \text{ and } \kappa_1 \to 0+; \tag{5.3.14}$$

and, in two dimensions,

$$G(\mathbf{r}) \sim B_2 (\ln r) \exp(-\kappa_1 r) \qquad \text{for fixed } r \text{ and } \kappa_1 \to 0+. \tag{5.3.15}$$

It is interesting to note that in the special case of three dimensions the asymptotic results (5.3.13) and (5.3.14) are identical and that, for $d \geqslant 3$, correlations at the critical point have the asymptotic power-law decay

$$G(\mathbf{r}) \sim B_d r^{-(d-2)} \qquad \text{at } T = T_c \text{ and } r \to \infty. \tag{5.3.16}$$

The two-dimensional result given in eqn (5.3.15), assuming there is a phase-transition point (where $\kappa_1 = 0$), is clearly nonsense since it predicts a logarithmically divergent correlation function at the critical point.

In the following section, we discuss a general form for the decay of correlation functions expressed in terms of certain critical exponents.

5.4 Correlation-function critical exponents

In his discussions on critical point fluctuations and correlations, Fisher (1962, 1964*b*) proposed a modification of the Ornstein–Zernike result (5.3.16) by assuming a power-law decay at the critical point characterized by the critical exponent η defined, in general, for d-dimensional systems by

$$G(\mathbf{r}) \sim B_d r^{-(d-2+\eta)} \qquad \text{at } T = T_c \text{ and } r \to \infty \tag{5.4.1}$$

and restricted by $0 \leqslant \eta \leqslant 2$. This assumption is certainly borne out by experiment and also theoretically by certain results for model lattice systems. In particular, the exact result obtained by Kaufman and Onsager (1949) for the two-dimensional Ising model is $\eta = 1/4$.

Also, by analogy with the results of the previous section, it is natural to *define* the inverse range of correlations κ_1 by

$$\kappa_1 = -\lim_{r \to \infty} r^{-1} \ln G(\mathbf{r}) \tag{5.4.2}$$

and its associated critical exponents ν and ν' by

$$\kappa_1 \sim \begin{cases} C_+\left[\dfrac{T}{T_c}-1\right]^{\nu} & \text{as } T\to T_c+ \\ C_-\left[1-\dfrac{T}{T_c}\right]^{\nu'} & \text{as } T\to T_c-. \end{cases} \tag{5.4.3}$$

In the event that the limit in eqn (5.4.2) does not exist there are other means of defining a correlation length (see, for example, Fisher and Burford 1967). For simplicity, we will assume exponential decay of correlation functions with correlation length ξ defined in terms of eqn (5.4.2) by $\xi=\kappa_1^{-1}$. The exponents ν and ν' defined by eqn (5.4.3) then describe the rate of divergence of the correlation length as the critical point is approached from above and below, respectively.

It will be noted that, when $G(\mathbf{r})$ has the general power-law decay given in eqn (5.4.1), its Fourier transform has the form

$$\hat{G}(\mathbf{q}) \sim \hat{B}q^{\eta-2} \qquad \text{at } T=T_c \quad \text{as } q\to 0. \tag{5.4.4}$$

Solving eqn (5.3.4) for $\hat{C}(\mathbf{q})$ in terms of $\hat{G}(\mathbf{q})$ then shows that, for fixed density,

$$\hat{C}(\mathbf{q}) \sim \rho[1-a^2q^{2-\eta}+\ldots] \qquad \text{at } T=T_c \quad \text{as } q\to 0. \tag{5.4.5}$$

It follows that, when $\eta>0$, the second moment of $C(\mathbf{r})$ does not exist and, moreover, on inverting eqn (5.4.5), that $C(\mathbf{r})$ has the long-range power-law decay

$$C(\mathbf{r}) \sim Dr^{-(d+2-\eta)} \qquad \text{as } r\to\infty. \tag{5.4.6}$$

Notice, however, that when $0<\eta<2$, $C(\mathbf{r})$ decays more rapidly than $G(\mathbf{r})$, particularly when η is small.

Classical values for the correlation-function critical exponents defined above are

$$\nu=\nu'=1/2, \qquad \eta=0. \tag{5.4.7}$$

The values for ν and ν', in particular, were derived from eqn (5.3.9) assuming the classical values of $\gamma=\gamma'=1$ for the compressibility exponents. More generally, if we don't assume classical behaviour for K_T, eqn (5.3.9) predicts

$$\nu=\gamma/2 \qquad \text{and} \qquad \nu'=\gamma'/2. \tag{5.4.8}$$

These results, however, rely on the Ornstein–Zernike assumptions about $C(\mathbf{r})$ which in turn imply that $\eta=0$.

If we return to the fluctuation relation given in eqn (5.2.10) and assume by analogy with eqns (5.3.14), (5.3.16), and (5.4.1) that

$$G(\mathbf{r}) \sim Br^{-(d-2+\eta)}\exp(-\kappa_1 r) \qquad \text{as } T\to T_c, \tag{5.4.9}$$

we have

$$\rho^2 kTK_T \sim A \int G(\mathbf{r}) r^{d-1}\,\mathrm{d}r$$
$$\sim D \int \exp(-\kappa_1 r) r^{1-\eta}\,\mathrm{d}r \qquad \text{as } T \to T_c \tag{5.4.10}$$

where A and D are constants. Changing variables to $x = \kappa_1 r$ then shows that, for fixed density,

$$K_T \sim E\kappa_1^{\eta-2} \qquad \text{as } T \to T_c. \tag{5.4.11}$$

From the definition (5.4.3) of the exponents ν and ν' and recalling that the corresponding exponents for K_T are γ and γ', we obtain the relations

$$\gamma = (2-\eta)\nu \qquad \text{and} \qquad \gamma' = (2-\eta)\nu'. \tag{5.4.12}$$

Since we have been rather casual in the above argument with integrals and limits our derivation of the exponent relations given in eqn (5.4.12) is certainly not rigorous. Nevertheless, these relations are valid for the two-dimensional Ising model (with $\gamma = \gamma' = 7/4$, $\eta = 1/4$, $\nu = \nu' = 1$) and also, more or less by definition, for the classical theories. Experimental results are consistent with eqns (5.4.12) although there are some uncertainties in estimates for η which range between 0 and about 0.1 with the classical value of $\eta = 0$ not being completely ruled out.

What is known rigorously (Fisher 1969) is that, under certain assumptions which we will not discuss here, we have the inequality

$$\gamma \leqslant (2-\eta)\nu. \tag{5.4.13}$$

Related inequalities due to Josephson (1967) are

$$(2-\alpha) \leqslant d\nu \qquad \text{and} \qquad (2-\alpha') \leqslant d\nu' \tag{5.4.14}$$

where α and α' are the specific heat exponents and d denotes the dimensionality of the system.

5.5 Spin correlations and the fluctuation relation for magnets

For simplicity, we will consider only Ising spin systems with interaction energy of the form

$$E\{\mu\} = -\sum_{1 \leqslant i < j \leqslant N} J_{ij}\mu_i\mu_j - H\sum_{i=1}^{N}\mu_i \tag{5.5.1}$$

where $\mu_i = \pm 1$, $i = 1, 2, \ldots, N$.

The average or expectation value of some functions f depending on configurations $\{\mu\}$ is defined by

$$\langle f\{\mu\}\rangle_N = Z(N, \beta, H)^{-1} \sum_{\{\mu\}} f\{\mu\} \exp(-\beta E\{\mu\}) \tag{5.5.2.}$$

where

$$Z(N, \beta, H) = \sum_{\{\mu\}} \exp(-\beta E\{\mu\}) \tag{5.5.3}$$

is the canonical partition function.

For example, the average magnetization per spin is given by

$$m_N(\beta, H) = \left\langle N^{-1} \sum_{i=1}^{N} \mu_i \right\rangle_N. \tag{5.5.4}$$

If the system is translationally invariant, that is, if

$$J_{ij} = J(\mathbf{r}_i - \mathbf{r}_j) \tag{5.5.5}$$

where $\mathbf{r}_i$ denotes the lattice vector for site i, and we assume periodic boundary conditions, $\langle \mu_i \rangle_N$ is independent of i and hence

$$m_N(\beta, H) = \langle \mu_1 \rangle_N \tag{5.5.6}$$

where the subscript '1' denotes an arbitrary site of the lattice. The spontaneous magnetization in this case is then given by

$$m_0(\beta) = \lim_{H \to 0+} \lim_{N \to \infty} m_N(\beta, H) = \langle \mu_1 \rangle \tag{5.5.7}$$

where, in general, the angular brackets without the subscript N denote the limit of eqn (5.5.2) as $N \to \infty$, assuming of course that the limit exists.

For our present purposes we define the *pair-correlation function* for an Ising spin system by

$$G_N(k, l) = \langle \mu_k \mu_l \rangle_N - \langle \mu_k \rangle_N \langle \mu_l \rangle_N. \tag{5.5.8}$$

If the system is translationally invariant, G_{kl} depends only on $\mathbf{r}_k - \mathbf{r}_l$ and in this case

$$G_N(k, l) = \langle \mu_k \mu_l \rangle_N - \langle \mu_1 \rangle_N^2. \tag{5.5.9}$$

Subsequently, we will use G to denote the value of G_N in the limit $N \to \infty$.

In zero field, $\langle \mu_k \rangle_N = 0$ by symmetry of $E\{\mu\}$ under spin reversal ($\mu_i \to -\mu_i$) so that the pair-correlation function in this case becomes simply $\langle \mu_k \mu_l \rangle_N$.

In a sense the expectation value $\langle \mu_k \mu_l \rangle$ is a measure of the degree of order in the lattice so that if spins are correlated over large distances one would expect

$$\lim_{|\mathbf{r}_k - \mathbf{r}_l| \to \infty} \langle \mu_k \mu_l \rangle \neq 0. \tag{5.5.10}$$

If eqn (5.5.10) is satisfied, we say that there is *long-range order* and in such a case one would expect the system to exhibit a phase transition. In fact, by analogy with the fluid case, one might expect the pair-correlation function $G(\mathbf{r})$ to decay to zero so that from eqn (5.5.9) one has

$$\lim_{|\mathbf{r}_k - \mathbf{r}_l| \to \infty} \langle \mu_k \mu_l \rangle = \langle \mu_1 \rangle^2. \tag{5.5.11}$$

If one now takes the field to zero, the right-hand side of eqn (5.5.11), from eqn (5.5.7), is the square of the spontaneous magnetization so that if eqn (5.5.11) is valid, long-range order exists in zero field if and only if the spontaneous magnetization is non-zero. There is, however, more to it than this since long-range order is usually defined in terms of the zero-field correlation function. That is,

$$\rho = \lim_{|\mathbf{r}_k - \mathbf{r}_l| \to \infty} \lim_{N \to \infty} \lim_{H \to 0+} \langle \mu_k \mu_l \rangle \tag{5.5.12}$$

is usually taken to be the measure of long-range order. Notice that if eqn (5.5.11) is valid and if the limits on N and H in eqn (5.5.12) can be reversed then indeed it follows that $\rho = m_0^2$. It is possible, however, that the inversion of limits is not valid, in which case one could conceivably have a non-zero spontaneous magnetization without long-range order.

In order to discuss the asymptotic behaviour of the pair-correlation function we now derive, by analogy with the fluid system, a fluctuation relation connecting the isothermal susceptibility $\chi_N(\beta, H)$ and the pair-correlation function.

From the definition of χ_N and eqns (5.5.1)–(5.5.4), we have that

$$\begin{aligned} \chi_N(\beta, H) &= \frac{\partial}{\partial H} \left\langle N^{-1} \sum_{i=1}^{N} \mu_i \right\rangle_N \\ &= \beta \left[\left\langle N^{-1} \left(\sum_{i=1}^{N} \mu_i \right)^2 \right\rangle_N - N^{-1} \left\langle \sum_{i=1}^{N} \mu_i \right\rangle_N^2 \right] \\ &= \beta N^{-1} \sum_{k,l=1}^{N} G_N(k, l) \end{aligned} \tag{5.5.13}$$

where, in the last step, we have used the definition of G_N. For a translationally invariant system we have from eqn (5.5.9) that

$$\chi_N(\beta, H) = \beta \sum_{k=1}^{N} (\langle \mu_1 \mu_k \rangle_N - \langle \mu_1 \rangle_N^2) \tag{5.5.14}$$

and hence, in the limit $N \to \infty$,

$$\chi(\beta, H) = \beta \sum_{\mathbf{r}} G(\mathbf{r}) \tag{5.5.15}$$

where the sum in eqn (5.5.15) is over the infinite lattice, and in deriving this result from eqn (5.5.14) we have chosen the spin μ_1 to be located at the origin.

The fluctuation relation for magnetic systems given in eqn (5.5.15) is to be compared with the corresponding relation for fluid systems given in eqn (5.2.10). In particular, it will be noted that, so long as χ is bounded, the lattice sum in eqn (5.5.15) must converge and hence

$$\lim_{r \to \infty} G(\mathbf{r}) = 0 \tag{5.5.16}$$

which then implies eqn (5.5.11).

The discussion in the previous sections applies equally well to magnetic systems and one can, in particular, define the correlation length exponents ν, ν', and η for magnetic systems (in zero field) in precisely the same way as we did in the previous section. Moreover, if one can freely interchange sums and limits ($N \to \infty$, $H \to 0+$) in eqns (5.5.13)–(5.5.15), the arguments in the previous section hold and one deduces that

$$\gamma = (2-\eta)\nu \qquad \text{and} \qquad \gamma' = (2-\eta)\nu'. \tag{5.5.17}$$

The justification of the interchange of limits for the zero-field two-dimensional Ising model with nearest-neighbour interactions only was given by Abraham (1972, 1973) so that, from the known results $\eta = 1/4$ and $\nu = \nu' = 1$, one can deduce the values $\gamma = \gamma' = 7/4$ for the susceptibility exponents. These values were in fact obtained heuristically by Fisher (1959) who in essence derived eqn (5.5.17) by an argument similar to the one presented in the previous section. We will have more to say about the critical behaviour of the Ising model in the following chapter.

5.6 Griffiths spin-correlation inequalities

Since the original work of Griffiths (1967) many inequalities relating pair and higher-order spin-correlation functions have been derived for continuous as well as discrete (Ising) spin systems. The underlying assumption in these inequalities is that the system is ferromagnetic. That is, the coupling constants between pairs or, more generally, groups of spins are non-negative. For simplicity, we will consider here only ferromagnetic Ising spin systems with pair interactions and discuss only the two inequalities derived originally by Griffiths. A nice review of generalized Griffiths-type inequalities can be found in Sylvester (1976).

Consider then an Ising system with interaction energy

$$E\{\mu\} = -\sum_{1 \leqslant i < j \leqslant N} J_{ij}\mu_i\mu_j, \qquad J_{ij} \geqslant 0 \tag{5.6.1}$$

where $\mu_i = \pm 1$, $i = 1, 2, \ldots, N$.

The first Griffiths inequality states that the expectation value of a product of any number of the spins is non-negative. That is

$$\langle \mu_{i_1}\mu_{i_2} \cdots \mu_{i_k} \rangle_N \geqslant 0 \tag{5.6.2}$$

where the expectation value is defined in eqn (5.5.2).

In its simplest form, which is all we will consider here, the second Griffiths inequality states that the pair-correlation function for an Ising spin system with interaction (5.6.1) is a monotonically increasing function of the coupling constants. That is

$$\frac{\partial}{\partial J_{kl}}\langle\mu_i\mu_j\rangle_N=\beta(\langle\mu_i\mu_j\mu_k\mu_l\rangle_N-\langle\mu_i\mu_j\rangle_N\langle\mu_k\mu_l\rangle_N)\geqslant 0. \qquad (5.6.3)$$

The first inequality (5.6.2) is more or less obvious and easy to prove, granted that the system is ferromagnetic. The second inequality (5.6.3) is intuitively obvious and, as we will see shortly, is almost as easy to prove as the first inequality.

The second inequality and its generalizations have many interesting consequences. For example, it is clear from eqn (5.6.3) that, by adding more ferromagnetic interactions, the pair-correlation function cannot decrease. In particular, if we increase the size (N) of the system by adding ferromagnetic bonds, the pair-correlation function, considered as a function of N, forms a monotonically increasing sequence. Since this sequence is bounded above ($\langle\mu_i\mu_j\rangle_N\leqslant\langle|\mu_i\mu_j|\rangle_N=1$) it follows that $\langle\mu_i\mu_j\rangle$ exists in the thermodynamic limit $N\to\infty$.

Another result which follows from eqn (5.6.3) is that if a d-dimensional ferromagnetic Ising system S_d has a phase transition with long-range order then the $(d+1)$-dimensional ferromagnetic Ising system S_{d+1}, whose component d-dimensional layers are S_d, also has a phase transition with long-range order. This is easily proved by noting that the pair-correlation function for S_d cannot be decreased when it is transformed into the corresponding function for S_{d+1} by coupling S_d-systems together with ferromagnetic bonds.

Since we know that the two-dimensional Ising ferromagnet with nearest-neighbour interactions has a phase transition with long-range order at low temperatures, we can conclude from the above observations that any two- or higher-dimensional Ising ferromagnet with at least nearest-neighbour interactions also has such a phase transition.

In order to prove the first inequality (5.6.2) we note from the definition of the expectation value that, since the partition function is non-negative, we need only prove that the numerator of the expectation value in eqn (5.6.2) is non-negative. That is, we need only show, from eqn (5.6.1), that

$$\sum_{\{\mu\}}(\mu_{i_1}\mu_{i_2}\cdots\mu_{i_k})\exp\Big(\beta\sum_{1\leqslant i<j\leqslant N}J_{ij}\mu_i\mu_j\Big)\geqslant 0. \qquad (5.6.4)$$

If we expand the exponential in eqn (5.6.4) we obtain, since $\beta J_{ij}\geqslant 0$, a multinomial in the spin variables with non-negative coefficients. The sum over spin configurations can then be performed independently per site with each site contributing terms of the form

$$\sum_{\mu_i=\pm 1}\mu_i^n=\begin{cases}0 & \text{when } n \text{ is odd}\\ 1 & \text{when } n \text{ is even.}\end{cases} \qquad (5.6.5)$$

The required result then follows immediately.

To prove the second Griffiths inequality (5.6.3) we use a clever idea, first introduced by Ginibre (1970), which considers a system together with its 'duplicate'. That is we consider, instead of the N-spin Ising system with interaction energy (5.6.1), the $2N$-spin Ising system composed of spins μ_i and μ'_i, $i=1, 2, \ldots, N$ with interaction energy, from eqn (5.6.1), given by

$$E\{\mu, \mu'\}=E\{\mu\}+E\{\mu'\}$$

$$=-\sum_{1\leqslant i<j\leqslant N} J_{ij}(\mu_i\mu_j+\mu'_i\mu'_j). \tag{5.6.6}$$

Denoting expectation values for the system eqn (5.6.6) by $\langle\cdot\rangle_{2N}$, it is clear that for arbitrary $f\{\mu\}$ and $g\{\mu\}$

$$\langle f\{\mu\}g\{\mu'\}\rangle_{2N}=\langle f\{\mu\}\rangle_N\langle g\{\mu\}\rangle_N. \tag{5.6.7}$$

It then follows from eqn (5.6.3) that Griffiths's second inequality is equivalent to showing that, in the duplicated system,

$$\langle\mu_i\mu_j(\mu_k\mu_l-\mu'_k\mu'_l)\rangle_{2N}\geqslant 0. \tag{5.6.8}$$

The next step is to introduce new spin variables σ_i and τ_i by the orthogonal change of variables

$$\sigma_i=\frac{1}{\sqrt{2}}(\mu_i+\mu'_i), \qquad \tau_i=\frac{1}{\sqrt{2}}(\mu_i-\mu'_i). \tag{5.6.9}$$

In terms of the σs and τs it is easily verified from eqn (5.6.6) that

$$E\{\mu, \mu'\}=-\sum_{1\leqslant i<j\leqslant N} J_{ij}(\sigma_i\sigma_j+\tau_i\tau_j) \tag{5.6.10}$$

and that the inequality in eqn (5.6.8) is equivalent to

$$\langle(\sigma_i+\tau_i)(\sigma_j+\tau_j)(\sigma_k\tau_l+\sigma_l\tau_k)\rangle_{2N}\geqslant 0. \tag{5.6.11}$$

If we now follow the argument given previously for Griffiths's first inequality and expand the Boltzmann factor

$$\exp(-\beta E\{\mu, \mu'\})=\exp\{\beta\sum_{1\leqslant i<j\leqslant N} J_{ij}(\sigma_i\sigma_j+\tau_i\tau_j)\} \tag{5.6.12}$$

in the numerator of the expectation value appearing in eqn (5.6.11), it is clear that we need only prove, by analogy with eqn (5.6.5), that for each site i and integers n and m,

$$\sum_{\mu_i=\pm 1,\, \mu'_i=\pm 1} \sigma_i^n\tau_i^m\geqslant 0. \tag{5.6.13}$$

The problem is further simplified for Ising spin systems for which

$$\sigma_i \tau_i = \tfrac{1}{2}(\mu_i^2 - \mu_i'^2) = 0. \tag{5.6.14}$$

It follows that our proof of Griffiths's second inequality will be complete once we have shown that, for an arbitrary integer n,

$$\sum_{\mu_i = \pm 1, \mu_i' = \pm 1} \sigma_i^n \geqslant 0 \quad \text{and} \quad \sum_{\mu_i = \pm 1, \mu_i' = \pm 1} \tau_i^n \geqslant 0. \tag{5.6.15}$$

By reversing the spins ($\mu_i \to -\mu_i$, $\mu_i' \to -\mu_i'$) in eqn (5.6.15) the summations are unaltered but, from eqn (5.6.9), the summands change sign when n is odd. In this case the inequalities in eqn (5.6.15) are satisfied as equalities whereas, when n is even, the sums in eqn (5.6.15) are positive by definition.

The proof of eqn (5.6.3) is then complete.

5.7 Problems

1. Show that the pair-correlation function $G(\mathbf{r})$ in the Ornstein–Zernike theory, whose Fourier transform is given by eqn (5.3.8), i.e.

$$(q^2 + \kappa_1^2) G(\mathbf{q}) = R^{-2} C(0) \equiv c$$

satisfies the partial differential equation

$$\nabla^2 G - \kappa_1^2 G = -c\delta(\mathbf{r})$$

where δ denotes the Dirac delta function.

2. Assuming that the Ornstein–Zernike pair-correlation function is a function only of $|\mathbf{r}| = r$, show that $G(r)$ for $r \neq 0$ satisfies the ordinary differential equation

$$\frac{\mathrm{d}^2 G}{\mathrm{d}r^2} + \frac{(d-1)}{r} \frac{\mathrm{d}G}{\mathrm{d}r} - \kappa_1^2 G = 0$$

where d is the dimensionality of the system.

Use this result to verify the asymptotic forms for $G(r)$ given in eqns (5.3.13)–(5.3.15).

3. Derive the asymptotic form given in eqn (5.4.6) for the direct correlation function $C(\mathbf{r})$.

4. The pair-correlation function for the zero-field one-dimensional Ising model with nearest-neighbour coupling constant J is given by

$$\langle \mu_k \mu_l \rangle = [\tanh \beta J]^{|k-l|}.$$

Use this result and the fluctuation relation (5.5.15) to obtain an expression for the zero-field isothermal susceptibility.

5. Compute the four-spin correlation function for the zero-field one-dimensional Ising model with nearest-neighbour interactions only and use eqn (5.6.3) to directly verify Griffiths' second correlation inequality.

5.8 Solutions to problems

1. Integrating twice by parts in each of the d spatial variables $x_1, \ldots, x_d$, gives

$$\int e^{i\mathbf{q}\cdot\mathbf{r}} \nabla^2 G \, d\mathbf{r} = \int_{-\infty}^{\infty} \cdots \int \exp\left(i \sum_{k=1}^{d} q_k x_k\right) \sum_{k=1}^{d} \frac{\partial^2 G}{\partial x_k^2} dx_1 \cdots dx_d$$

$$= -\left(\sum_{k=1}^{d} q_k^2\right) \int \exp(i\mathbf{q}\cdot\mathbf{r}) G(\mathbf{r}) d\mathbf{r}$$

$$= -q^2 \hat{G}(\mathbf{q})$$

$$= \kappa_1^2 \hat{G}(\mathbf{q}) + c$$

where, in the last step, we have used the Ornstein–Zernike expression for $\hat{G}(\mathbf{q})$.

It follows immediately that, for all $\mathbf{q}$,

$$\int e^{i\mathbf{q}\cdot\mathbf{r}} [\nabla^2 G - \kappa_1^2 G - c\delta(\mathbf{r})] d\mathbf{r} = 0$$

which gives the required result.

2. Assuming that $G(\mathbf{r}) = G(r)$ with $r = (x_1^2 + x_2^2 + \ldots + x_d^2)^{1/2}$, it is easily verified that

$$\nabla^2 G = \sum_{k=1}^{d} \frac{\partial^2 G}{\partial x_k^2} = \frac{d^2 G}{dr^2} + \frac{d-1}{r}\frac{dG}{dr}.$$

The required result then follows from Problem 1.

When

$$G(r) = r^{-\alpha} \exp(-\kappa_1 r),$$

$$\nabla^2 G - \kappa_1^2 G = \left[\frac{\alpha}{r^2}(\alpha + 2 - d) + \frac{\kappa_1}{r}(2\alpha - d + 1)\right] G(r).$$

The leading asymptotic form for $G(r)$ as $r \to \infty$ for κ_1 fixed is obtained by choosing $\alpha = (d-1)/2$ while, for $\kappa_1 \to 0$, it is appropriate to choose $\alpha = d - 2$.

Note that, when $d = 3$, both expressions give $\alpha = 1$ while, when $d = 2$, the choice $G(r) = \ln r \exp(-\kappa_1 r)$ gives

$$\nabla^2 G - \kappa_1^2 G = -\frac{\kappa_1}{r}(2 + \ln r) \exp(-\kappa_1 r) \sim 0 \qquad \text{as } \kappa_1 \to 0.$$

3. Taking the inverse Fourier transform and using polar coordinates we obtain

$$C(\mathbf{r}) = (2\pi)^{-d} \int \hat{C}(\mathbf{q}) e^{i\mathbf{q}\cdot\mathbf{r}} d\mathbf{q}$$

$$\sim C \int q^{2-\eta} e^{iqr\cos\theta} q^{d-1} dq d\Omega \qquad \text{as } r \to \infty.$$

Changing variable to $x = qr$ then gives

$$C(\mathbf{r}) \sim C r^{-(d+2-\eta)} \int x^{d+1-\eta} e^{-ix\cos\theta} dx \, d\Omega$$

which is the stated asymptotic form in eqn (5.4.6).

4. It is to be stressed that the sum in eqn (5.5.15) is over the 'infinite lattice'. In one dimension and in zero field we then have

$$\begin{aligned}\chi(\beta, 0) &= \beta \sum_{k=-\infty}^{\infty} (\tanh \beta J)^{|k|} \\ &= \beta[1+2 \tanh \beta J/(1-\tanh \beta J)] \\ &= \beta e^{2\beta J}.\end{aligned}$$

This result is to be compared with the thermodynamic definition

$$\chi(\beta, 0) = \lim_{H\to 0+} \frac{\partial m}{\partial H}$$

and the expression (6.3.9)

$$m(\beta, H) = \sinh \beta H/(\sinh^2 \beta H + e^{-4K})^{1/2}$$

for the magnetization of the one-dimensional Ising model with nearest-neighbour interactions.

5. Using a slight generalization of the argument given in Problem 8 of Chapter 3, it is not difficult to show that for $k > l > m > n$,

$$\langle \mu_k \mu_l \mu_m \mu_n \rangle = (\tanh \beta J)^{k-l+m-n}.$$

Griffiths's second correlation inequality is then easily verified by noting the three possibilities for eqn (5.6.3)

$$\langle \mu_k \mu_l \mu_m \mu_n \rangle - \langle \mu_k \mu_l \rangle \langle \mu_m \mu_n \rangle = 0$$

and, for $J \geqslant 0$,

$$\begin{aligned}\langle \mu_k \mu_l \mu_m \mu_n \rangle - \langle \mu_k \mu_m \rangle \langle \mu_l \mu_n \rangle &= \langle \mu_k \mu_l \mu_m \mu_n \rangle - \langle \mu_k \mu_n \rangle \langle \mu_l \mu_m \rangle \\ &= (\tanh \beta J)^{k-l+m-n}[1-(\tanh \beta J)^{2(l-m)}] \geqslant 0.\end{aligned}$$

6

SOME EXACTLY SOLVED MODELS

6.1 Introduction

In Chapter 3, we discussed several models which can be solved exactly. These were the one-dimensional Tonks and Takahashi gases and the one-dimensional zero-field n-vector model with nearest-neighbour interactions. Although these models can be solved exactly, they are uninteresting in the sense that their respective free energies in the thermodynamic limit are analytic functions of the appropriate intensive variables T and v or H. In other words, these models do not undergo phase transitions.

In Chapter 4, we studied the van der Waals and Curie–Weiss models and obtained explicit expressions for their free energies. These models exhibit singular behaviour and reproduce the classical theories of phase transitions but, as we saw, these theories are only strictly valid in the long-range or Kac limit. The models themselves are also unrealistic since the interactions depend explicitly on the size of the system. As we saw in Section 4.2, this unphysical feature results in the so-called van der Waals wiggle which violates the fundamental principles of thermodynamic stability.

As mentioned previously, the first realistic model which was solved exactly and which exhibited a phase transition was the two-dimensional Ising model with nearest-neighbour interactions in zero external field. The exact expression for the partition function of this model was first derived by Onsager (1944) in one of the most celebrated and much quoted papers in mathematical physics. The surprising feature of Onsager's solution at the time was that the model exhibited a phase transition with a symmetric logarithmic divergence in the specific heat, which was, of course, at variance with the predictions of the classical theories. It was known from the work of Peierls (1936*a*), which we discussed in Section 4.7, that the model had a phase transition, but prior to Onsager, various approximation schemes due to Bragg and Williams (1934, 1935), Bethe (1935), Peierls (1936*a*, *b*), and others all predicted classical critical behaviour. One of the purposes of trying to find exact solutions in equilibrium statistical mechanics is, of course, to provide a testing ground for approximate methods and in so doing one hopes that some insight will be gained into the intricacies of phase transitions.

Since Onsager's solution of the two-dimensional Ising model in zero field, numerous man-hours of effort and many a good researcher falling foul to the dreaded 'Ising's disease' have failed to yield exact solutions to the two-dimensional model in non-zero field and the three-dimensional model, even in zero field. However, another important class of lattice models, loosely termed vertex models, have been solved exactly in two dimensions. These include the 'ice', 'F', and 'KDP' models solved by Lieb (1967*a*, *b*, *c*) and the now famous solution of the eight-vertex model due to Baxter (1971, 1972). Most recently, Baxter (1980) has added the hard-hexagon model, which can be thought of as a lattice gas on a triangular lattice with nearest-neighbour exclusion, to the list of exactly solved models.

Since Baxter (1982) has written an extensive and detailed review of exactly solved lattice models, we will present here only the essential features and results for the two-dimensional Ising, vertex, and hard-hexagon models. The transfer-matrix method, which forms the basis for the solution of all these models, is discussed in the following section.

In addition to the lattice models which have only been solved exactly in two dimensions we conclude this chapter with a discussion of the Gaussian and spherical models which can be solved exactly in any dimension.

6.2 The transfer-matrix method for lattice systems

The original idea of the transfer-matrix method was due to Kramers and Wannier (1941) and it, in fact, formed the basis for Onsager's solution of the two-dimensional model. It has since proved useful in formulating and analysing a number of model systems in statistical mechanics, some of which we will consider in this chapter.

In a general setting let us suppose that we have a lattice composed of N 'layers', which could, for example, consist of points, lines, or planes for one-, two-, and three-dimensional lattices, respectively. If we suppose that the 'state' of the ith layer is specified by a quantity σ_i, which can take only a finite number of discrete values, and that interactions occur only within layers and between nearest-neighbour layers, the interaction energy $E\{\sigma\}$ in a given configuration $\{\sigma\}=(\sigma_1, \sigma_2, \ldots, \sigma_N)$ or state of the lattice can, in general, be expressed in the form

$$E\{\sigma\}=\sum_{i=1}^{N} [U(\sigma_i)+V(\sigma_i, \sigma_{i+1})] \tag{6.2.1}$$

where $U(\sigma_i)$ represents the interaction energy within a layer in state σ_i and $V(\sigma_i, \sigma_{i+1})$ represents the interaction energy between nearest-neighbour layers in states σ_i and σ_{i+1}, respectively, and for simplicity we have assumed periodic boundary conditions. That is, we assume that the Nth layer is coupled to the first or equivalently, that $\sigma_{N+1}\equiv\sigma_1$.

The partition function for a lattice model whose interaction energy can be expressed in the form (6.2.1) is given by

$$Z_N = \sum_{\{\sigma\}} \exp(-\beta E\{\mu\})$$
$$= \sum_{(\sigma_1, \ldots, \sigma_N)} \exp\left[-\beta \sum_{i=1}^{N} \{U(\sigma_i) + V(\sigma_i, \sigma_{i+1})\}\right]$$
$$= \sum_{(\sigma_1, \ldots, \sigma_N)} \prod_{i=1}^{N} T(\sigma_i, \sigma_{i+1}) \tag{6.2.2}$$

where, by symmetrizing the exponent,

$$T(\sigma, \sigma') = \exp[-\beta\{U(\sigma)/2 + V(\sigma, \sigma') + U(\sigma')/2\}]. \tag{6.2.3}$$

If we now consider the quantities $T(\sigma, \sigma')$ as components of an $M \times M$ matrix $\mathbb{T}$, where M is the number of possible values for each layer state σ, it will be noted that the product over i in eqn (6.2.2) resembles a matrix product. Thus, if we sum over states σ_2 in eqn (6.2.2), for example, we have

$$\sum_{\sigma_2} T(\sigma_1, \sigma_2) T(\sigma_2, \sigma_3) = \mathbb{T}^2(\sigma_1, \sigma_3) \tag{6.2.4}$$

where $\mathbb{T}^2(\sigma_1, \sigma_3)$ denotes the (σ_1, σ_3) component of the matrix $\mathbb{T}^2$. It will be noted that the matrix $\mathbb{T}$ in effect transfers us from one layer to the next, hence the term transfer matrix.

Repeating the above process and summing successively over layer states $\sigma_2, \ldots, \sigma_N$ in eqn (6.2.2) we then obtain

$$Z_N = \sum_{\sigma_1} \mathbb{T}^N(\sigma_1, \sigma_1)$$
$$= \mathrm{tr}(\mathbb{T}^N) \tag{6.2.5}$$

where 'tr' denotes the trace of the matrix, i.e. the sum of the diagonal entries.

Using now the fact that the trace of a matrix is also the sum of its eigenvalues and that the eigenvalues of $\mathbb{T}^N$ are λ_i^N, $i = 1, 2, \ldots, M$ where λ_i, $i = 1, 2, \ldots, M$ are the eigenvalues of $\mathbb{T}$ eqn (6.2.5) shows that

$$Z_N = \sum_{i=1}^{M} \lambda_i^N. \tag{6.2.6}$$

Now since the elements $T(\sigma, \sigma')$ are all positive it follows from the Perron–Frobenius theorem (see Gantmacher 1964) that the maximum eigenvalue is non-degenerate and an analytic function of its arguments and, in particular, from eqn (6.2.3), an analytic function of β or the temperature $T\,(\neq 0)$. Without loss of generality, we can then order the eigenvalues such that

$$\lambda_1 > |\lambda_2| \geqslant |\lambda_3| \geqslant \ldots \geqslant |\lambda_M| \tag{6.2.7}$$

and take λ_1 to be an analytic function of β.

If we now take a (partial) thermodynamic limit by allowing the number of layers $N \to \infty$ we obtain from eqns (6.2.6) and (6.2.7)

$$\begin{aligned} -\beta\psi_M &= \lim_{N\to\infty} N^{-1} \ln Z_N \\ &= \lim_{N\to\infty} N^{-1} \ln \left[\lambda_1^N \left(1 + \sum_{i=2}^{M} (\lambda_2/\lambda_1)^N \right) \right] \\ &= \ln \lambda_1 + \lim_{N\to\infty} N^{-1} \ln \left[1 + \sum_{i=2}^{M} (\lambda_2/\lambda_1)^N \right] \\ &= \ln \lambda_1 \end{aligned} \tag{6.2.8}$$

where ψ_M denotes the free energy per layer of the infinite layer system.

For systems whose dimensionality is two or more one needs to take the layer size to infinity to complete the thermodynamic limit. This means that one further limiting process ($M \to \infty$) is required. In any event we have in all cases that the limiting free energy is simply expressed by eqn (6.2.8) in terms of the maximum eigenvalue λ_1 of the transfer matrix and that for infinite-by-finite layer systems the free energy ψ_M is an analytic function of temperature. Such quasi one-dimensional systems, therefore, have no phase transition.

In the following section, we illustrate the transfer-matrix method by evaluating the limiting free energy for the one-dimensional Ising model with nearest-neighbour interactions in an external magnetic field.

We note, in conclusion, that the surface terms in eqn (6.2.1) arising from periodic boundary conditions make no contribution in the limit given in eqn (6.2.8). ψ_M defined by eqn (6.2.8) is then independent of the boundary conditions.

6.3 The one-dimensional Ising model in a magnetic field

The interaction energy for a chain of N Ising spins $\mu_i = \pm 1$, $i = 1, 2, \ldots, N$ with nearest-neighbour interactions only and in the presence of a magnetic field H is given, in configuration $\{\mu\} = (\mu_1, \mu_2, \ldots, \mu_N)$, by

$$E\{\mu\} = -J \sum_{i=1}^{N} \mu_i \mu_{i+1} - H \sum_{i=1}^{N} \mu_i \tag{6.3.1}$$

where J is the nearest-neighbour coupling constant and, as in eqn (6.2.1), we have imposed periodic boundary conditions by coupling the last spin to the first, or equivalently, set $\mu_{N+1} \equiv \mu_1$.

In the notation of the previous section, the layers are simply lattice sites on the chain, the number of states per layer (M) is two, corresponding to spin up ($\mu = +1$) or spin down ($\mu = -1$), the interaction energy between nearest-neighbour layers is given by

$$V(\mu_i, \mu_{i+1}) = -J \mu_i \mu_{i+1} \tag{6.3.2}$$

and the interaction energy within each layer by

$$U(\mu_i) = -H\mu_i. \tag{6.3.3}$$

The transfer matrix $\mathbb{T}$ is then a 2×2 matrix with entries given from eqn (6.2.3) by

$$T(\mu, \mu') = \exp\left[K\mu\mu' + \frac{B}{2}(\mu+\mu')\right] \tag{6.3.4}$$

where $K = \beta J$ and $B = \beta H$.

Written out explicitly,

$$\mathbb{T} = \begin{pmatrix} T(+1, +1) & T(+1, -1) \\ T(-1, +1) & T(-1, -1) \end{pmatrix}$$
$$= \begin{pmatrix} \exp(K+B) & \exp(-K) \\ \exp(-K) & \exp(K-B) \end{pmatrix} \tag{6.3.5}$$

The eigenvalues of $\mathbb{T}$ are easily found to be

$$\left.\begin{matrix}\lambda_1 \\ \lambda_2\end{matrix}\right\} = e^K \cosh B \pm (e^{2K}\sinh^2 B + e^{-2K})^{1/2} \tag{6.3.6}$$

and hence, from eqn (6.2.8), the limiting free energy per spin $\psi(\beta, H)$ is given by

$$-\beta\psi(\beta, H) = \ln \lambda_1$$
$$= \ln\left[e^K \cosh B + (e^{2K}\sinh^2 B + e^{-2K})^{1/2}\right]. \tag{6.3.7}$$

It will be noted that in zero field ($B = 0$) this expression reduces to

$$-\beta\psi(\beta, 0) = \ln(2\cosh K) \tag{6.3.8}$$

which was the result obtained in Section 3.5 for the open chain.

Finally, we note that the magnetization is given by

$$m(\beta, H) = -\frac{\partial\psi}{\partial H}$$
$$= \sinh B/(\sinh^2 B + e^{-4K})^{1/2}. \tag{6.3.9}$$

It follows, as expected, that the spontaneous magnetization

$$m_0 = \lim_{H\to 0+} m(\beta, H) = 0 \qquad \text{for all } \beta > 0. \tag{6.3.10}$$

Notice, however, that at zero temperature ($K = \beta J \to \infty$) the magnetization is unity, so that if one stretches the point one could say that the one-dimensional Ising model with nearest-neighbour interactions has a phase transition at zero temperature.

6.4 Onsager's solution of the two-dimensional Ising model

For the two-dimensional Ising model on a rectangular lattice consisting of m rows and n columns with spins $\mu_{i,j}$, $i=1,2,\ldots,m$ and $j=1,2,\ldots,n$ located at the intersection of the ith row and jth column, the interaction energy, assuming nearest–neighbour interactions only, can be written as

$$E\{\mu\}=-J\sum_{i=1}^{m-1}\sum_{j=1}^{n}\mu_{i,j}\mu_{i+1,j}-J'\sum_{i=1}^{m}\sum_{j=1}^{n}\mu_{i,j}\mu_{i,j+1}-H\sum_{i=1}^{m}\sum_{j=1}^{n}\mu_{i,j} \tag{6.4.1}$$

where J and J' are the nearest-neighbour coupling constants for spins in columns and rows, respectively, and again for simplicity we have imposed periodic boundary conditions $\mu_{i,n+1}\equiv\mu_{i,1}$, $i=1,2,\ldots,m$ by what effectively amounts to wrapping the lattice around a cylinder.

In the notation of Section 6.2, the layers are columns; the jth layer having state

$$\sigma_j=(\mu_{1,j},\mu_{2,j},\ldots,\mu_{m,j}). \tag{6.4.2}$$

The interaction energy of the jth layer is given by

$$U(\sigma_j)=-J\sum_{i=1}^{m-1}\mu_{i,j}\mu_{i+1,j}-H\sum_{i=1}^{m}\mu_{i,j} \tag{6.4.3}$$

and the interaction energy between nearest-neighbour layers is given by

$$V(\sigma_j,\sigma_{j+1})=-J'\sum_{i=1}^{m}\mu_{i,j}\mu_{i,j+1}. \tag{6.4.4}$$

The total interaction energy (6.4.1) is then of the form (6.2.1) and hence the transfer matrix has entries given by eqn (6.2.3) which, from eqns (6.4.3) and (6.4.4) have the form

$$\begin{aligned}T(\sigma,\sigma')&=\exp[-\beta U(\sigma)/2]\exp[-\beta V(\sigma,\sigma')]\exp[-\beta U(\sigma')/2]\\&=\exp\left[\frac{K}{2}\sum_{i=1}^{m-1}\mu_i\mu_{i+1}+\frac{B}{2}\sum_{i=1}^{m}\mu_i\right]\exp\left[K'\sum_{i=1}^{m}\mu_i\mu_i'\right]\\&\quad\times\exp\left[\frac{K}{2}\sum_{i=1}^{m-1}\mu_i'\mu_{i+1}'+\frac{B}{2}\sum_{i=1}^{m}\mu_i'\right]\end{aligned} \tag{6.4.5}$$

where $\sigma=(\mu_1,\mu_2,\ldots,\mu_m)$, $\sigma'=(\mu_1',\mu_2',\ldots,\mu_m')$, $K=\beta J$, $K'=\beta J'$, and $B=\beta H$.

Note that the number of layer states, M, in our previous notation, is simply the number of column configurations given in eqn (6.4.2), i.e. $M=2^m$ since each spin in a column can take two values. In the transfer-matrix formalism the effect of going from one to two dimensions is then quite dramatic. Thus, in one dimension we had merely to diagonalize a 2×2 matrix but in two dimensions we have to diagonalize a $2^m\times2^m$ matrix, and then compute the largest eigenvalue in the limit $m\to\infty$.

If we denote the partition function for our rectangular lattice of nm spins by $Z_{n,m}$ the free energy in the thermodynamic limit is given from eqn (6.2.8) by

$$-\beta\psi(\beta, H) = \lim_{m,n\to\infty} (mn)^{-1} Z_{n,m}$$
$$= \lim_{m\to\infty} m^{-1} \ln \lambda_1. \tag{6.4.6}$$

Onsager (1944) actually obtained the complete spectrum of the transfer matrix for $H=0$ and arbitrary K and K'. For simplicity, we will consider here the square-lattice case $K=K'$ and consider only the limiting eqn (6.4.6).

The expression obtained by Onsager (1944) for the maximum eigenvalue of $\mathbb{T}$ (6.4.5) is

$$\lambda_1 = (2\sinh 2K)^{m/2} \exp[\tfrac{1}{2}(\gamma_1 + \gamma_3 + \ldots + \gamma_{2m-1})] \tag{6.4.7}$$

where γ_k is defined by

$$\cosh\gamma_k = \cosh 2K \coth 2K - \cos(\pi k/m). \tag{6.4.8}$$

The exponent in eqn (6.4.7) is a Riemann sum which approaches an integral in the limit $m\to\infty$, yielding from eqns (6.4.6) and (6.4.8) the result

$$-\beta\psi(\beta, 0) = \frac{1}{2}\ln(2\sinh 2K) + \lim_{m\to\infty}(2m)^{-1}\sum_{k=0}^{m-1}\gamma_{2k+1}$$
$$= \frac{1}{2}\ln(2\sinh 2K) + (2\pi)^{-1}\int_0^\pi \operatorname{arcosh}(\cosh 2K\coth 2K - \cos\theta_1)\mathrm{d}\theta_1$$
$$= \ln 2 + (2\pi^2)^{-1}\int\int_0^\pi \ln[\cosh^2 2K - \sinh 2K(\cos\theta_1$$
$$+\cos\theta_2)]\mathrm{d}\theta_1\mathrm{d}\theta_2 \tag{6.4.9}$$

where, in the last step, we have used the identity

$$\operatorname{arcosh}|z| = \pi^{-1}\int_0^\pi \ln[2(z-\cos\theta_2)]\mathrm{d}\theta_2. \tag{6.4.10}$$

From eqn (6.4.9) the internal energy E is given by

$$E = \frac{\partial}{\partial\beta}(\beta\psi)$$
$$= -J\coth 2K\left[1 + (\sinh^2 2K - 1)\pi^{-2}\int\int_0^\pi \{\cosh^2 2K\right.$$
$$\left.-\sinh 2K(\cos\theta_1 + \cos\theta_2)\}^{-1}\mathrm{d}\theta_1\mathrm{d}\theta_2\right]. \tag{6.4.11}$$

The integral appearing in eqn (6.4.11) diverges logarithmically when $\cosh^2 2K = 2\sinh 2K$. To see how this comes about, note that in the neighbourhood of the origin $\theta_1=\theta_2=0$, when

$$\delta=\cosh^2 2K-2\sinh 2K$$
$$=(\sinh 2K-1)^2 \tag{6.4.12}$$

is small,

$$\pi^{-2}\iint_0^\pi \{\cosh^2 2K-\sinh 2K(\cos\theta_1+\cos\theta_2)\}^{-1}\mathrm{d}\theta_1\mathrm{d}\theta_2$$
$$\sim\pi^{-2}\iint_0 \left\{\delta+\frac{1}{2}\sinh 2K(\theta_1^2+\theta_2^2)\right\}^{-1}\mathrm{d}\theta_1\mathrm{d}\theta_2$$
$$=\frac{2}{\pi}\int_0\left(\delta+\frac{r^2}{2}\sinh 2K\right)^{-1}r\,\mathrm{d}r$$
$$\sim -2\pi^{-1}\ln\delta \qquad \text{as } \delta\to 0 \tag{6.4.13}$$

where, in the second to last step, we have transformed to polar coordinates so that $\theta_1^2+\theta_2^2=r^2$, $\mathrm{d}\theta_1\mathrm{d}\theta_2=r\mathrm{d}r\ \mathrm{d}\theta$ and so forth.

It follows, from eqn (6.4.13), that there is a singularity or a phase transition when $\delta=0$, i.e. from eqn (6.4.12) at $K=K_c$ given by

$$\sinh 2K_c=1. \tag{6.4.14}$$

Combining eqns (6.4.11)–(6.4.14), we see that E is continuous at the critical point and that, in the neighbourhood of K_c,

$$E\sim -J\coth 2K_c(1+A(K-K_c)\ln|K-K_c|) \tag{6.4.15}$$

where A is some constant. It follows from this result that the zero-field specific heat

$$C_0=\frac{\partial E}{\partial T}\sim B\ln|K-K_c| \qquad \text{as } K\to K_c \tag{6.4.16}$$

has a symmetric logarithmic divergence at the critical point.

Closed-form expressions for E and C_0 can, in fact, be given in terms of complete elliptic integrals. These are, respectively,

$$E=-J\coth 2K\{1+(2\tanh^2 2K-1)\frac{2}{\pi}K(k_1)\} \tag{6.4.17}$$

where

$$k_1=2\sinh 2K/\cosh^2 2K \tag{6.4.18}$$

and $K(k_1)$ is the complete elliptic integral of the first kind defined by

$$K(k_1)=\int_0^{\pi/2}(1-k_1^2\sin^2\theta)^{-1/2}\mathrm{d}\theta \tag{6.4.19}$$

and

$$C_0 = \frac{2k}{\pi}(K \coth 2K)^2 \Bigg[2K(k_1) - 2E(k_1)$$

$$- 2(1 - \tanh^2 2K) \left\{ \frac{\pi}{2} + (2 \tanh^2 2K - 1) K(k_1) \right\} \Bigg] \tag{6.4.20}$$

where $E(k_1)$ is the complete elliptic integral of the second kind defined by

$$E(k_1) = \int_0^{\pi/2} (1 - k_1^2 \sin^2 \theta)^{1/2} \, d\theta. \tag{6.4.21}$$

The critical point eqn (6.4.14) corresponds to $k_1 = 1$ and, in the neighbourhood of $k_1 = 1-$,

$$K(k_1) \sim \ln \{4(1 - k_1^2)^{-1/2}\} \tag{6.4.22}$$

which accounts for the 'Onsager logarithm' in the specific heat.

Since Onsager there have been numerous derivations of his famous formula (6.4.9) for the free energy (e.g. Kaufman 1949; Kac and Ward 1952; Schultz *et al.* 1964; Vdovichenko 1965; Thompson 1965; Glasser 1970; Baxter and Enting 1978, to mention just a few). All methods, however, fail to generalize to the case of non-zero field in two dimensions or to three dimensions even in zero field.

6.5 Critical behaviour of the two-dimensional Ising model

As we have seen in the previous section, the zero-field specific heat for the two-dimensional Ising model has a symmetric logarithmic divergence at the critical point. In terms of critical exponents defined in Section 4.5 we thus have

$$\alpha = \alpha' = 0_{\ln}. \tag{6.5.1}$$

The calculation of the spontaneous magnetization

$$m_0 = - \lim_{H \to 0+} \frac{\partial}{\partial H} \psi(\beta, H) \tag{6.5.2}$$

would appear to require the evaluation of $\psi(\beta, H)$ which has not to date been done. However, by various indirect methods it is known that m_0 for the two-dimensional Ising model has the remarkably simple form

$$m_0 = \begin{cases} \{1 - (\sinh 2K)^{-4}\}^{1/8} & \text{when } T < T_c \\ 0 & \text{when } T > T_c \end{cases} \tag{6.5.3}$$

yielding the critical exponent

$$\beta = 1/8. \tag{6.5.4}$$

The expression given in eqn (6.5.3) was first derived in the mid-1940s and 'announced' by Onsager (1949) but in typical Onsager fashion the details were never published. A brief account, however, along with some interesting historical material, can be found in a rare article written by Onsager (1971). The first published derivation of the Onsager cryptogram (6.5.3) was given by Yang (1952) and was extremely complicated. Simpler derivations have been given subsequently by Montroll *et al.* (1963) and by Baxter (1981) but the simplicity of the final result is still somewhat of a surprise.

No exact expression has been obtained for the zero-field susceptibility but it has been 'known' for a long time that

$$\chi_0 \sim A_{\pm}\left|1-\frac{T}{T_c}\right|^{-7/4} \qquad \text{as } T \to T_c\pm. \tag{6.5.5}$$

That is, in terms of critical exponents,

$$\gamma = \gamma' = 7/4. \tag{6.5.6}$$

These values were first suggested by Domb and Sykes (1957) on the basis of series analysis and have since been rigorously established by Abraham (1972, 1973). A little later, Barouch *et al.* (1973) calculated the amplitudes $A_{\pm}$ for the high- and low-temperature divergence of χ_0. Similarly, the critical isotherm is known to have the form

$$H \sim m^{15} \qquad \text{at } T = T_c \text{ as } H \to 0 \tag{6.5.7}$$

and hence, we have the critical exponent

$$\delta = 15. \tag{6.5.8}$$

It will be noted that the inequalities for critical exponents discussed in Section 4.5 are satisfied as equalities for the two-dimensional Ising model. That is, from eqns (6.5.1), (6.5.4), (6.5.6), and (6.5.8), we have

$$\alpha' + 2\beta + \gamma' = 2 \qquad \text{and} \qquad \alpha' + \beta(1+\delta) = 2. \tag{6.5.9}$$

The form of the pair-correlation function was first studied by Onsager (1944) who showed, among other things, that for $T > T_c$

$$\langle \mu_0 \mu_r \rangle \sim w_r(T)\exp(-\kappa(T)r) \qquad \text{as } r \to \infty \tag{6.5.10}$$

where the *inverse correlation length*

$$\kappa(T) \sim c(T - T_c) \qquad \text{as } T \to T_c + \tag{6.5.11}$$

has the critical exponent

$$\nu = 1. \tag{6.5.12}$$

Subsequently it was shown by Kaufman and Onsager (1949) that, when

$\mathbf{r}=(l, l)$ and $T=T_c$,

$$\langle \mu_0 \mu_{\mathbf{r}} \rangle \sim A r^{-1/4} \qquad \text{as } r \to \infty \tag{6.5.13}$$

which gives the critical exponent

$$\eta = 1/4. \tag{6.5.14}$$

The above results are in fact still valid for non-diagonal $\mathbf{r}$ but then the constants A in eqn (6.5.13) and c in eqn (6.5.11) have a slight angular dependence.

With exponents ν and η added to our list it will be noted that the two relations

$$d\nu = 2-\alpha, \qquad \gamma = (2-\eta)\nu \tag{6.5.15}$$

discussed in Section 5.4 are also valid. The second relation in eqn (6.5.15) was derived from the fluctuation relation which in fact formed the basis of the proof of eqn (6.5.6) given by Abraham (1972, 1973), and an earlier heuristic argument by Fisher (1959) which supported the estimate $\gamma = 7/4$.

Finally, it should be mentioned that the above values for the various critical exponents are also valid for any regular planar two-dimensional Ising model with nearest-neighbour interactions. The only things that vary from lattice to lattice in this case are the location of the critical point and the value of various critical-point amplitudes of singular thermodynamic quantities.

6.6 Vertex models in two dimensions

The motivation for the models to be discussed in this section came from various hydrogen-bonded crystals which exhibit phase transitions. Funnily enough, it appears to be Onsager (1939) who first suggested that the ferroelectric transition in the hydrogen-bonded crystal KH_2PO_4 is connected with an ordering of the hydrogen atoms in the bonds.

Onsager, in fact, had a lifelong devotion to ice which is the simplest system to be modelled as a vertex model. Specifically, if one imagines ice to be embedded in a regular square lattice with oxygen atoms sitting on the vertices of the lattice, we can model ice by assuming:

1. There is one and only one hydrogen atom on each bond.
2. There are exactly two hydrogen atoms near each vertex.

This model is due to Slater (1941) and was motivated by previous work of Bernal and Fowler (1933) and Pauling (1935).

Assumption 2 above is called the '*ice rule*' and implies that there are only six possible 'vertex configurations' as shown in Fig. 6.1, with solid circles denoting hydrogen atoms close to a vertex and crosses denoting hydrogen atoms away from a vertex.

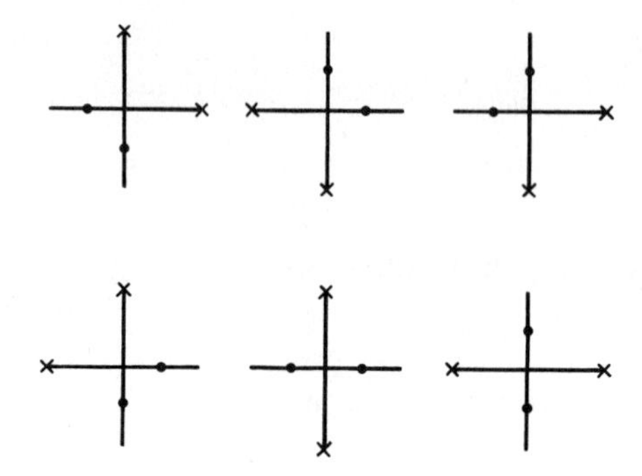

FIG. 6.1 Ice configurations.

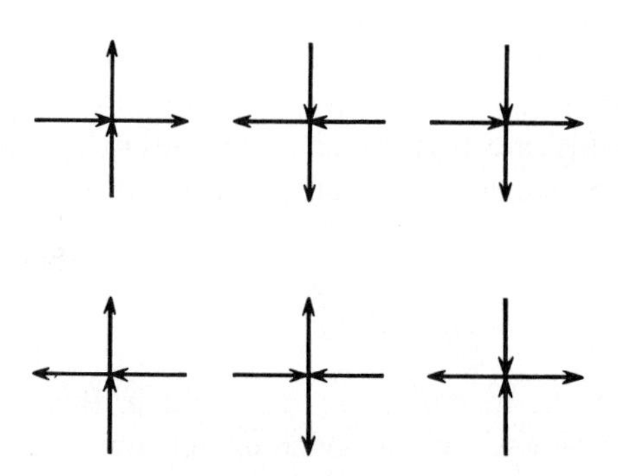

FIG. 6.2 The equivalent six-vertex model.

Equivalently, in Fig. 6.2, one can represent a hydrogen atom close to a vertex by an arrow on the bond pointing towards the vertex, and a hydrogen atom far from a vertex by an arrow pointing away from the vertex. This is the so-called *six-vertex model* which is obviously equivalent to the ice model shown in Fig. 6.1. We note, in particular, that the ice rule in terms of arrows on the bonds states that the number of arrows pointing towards a vertex is equal to the number of arrows pointing away from that vertex.

The simplest problem here is to calculate the residual entropy of square ice defined by

$$S = k \ln Z_0 \tag{6.6.1}$$

where Z_0 is the number of allowed configurations on the lattice that are consistent with the ice rule. Various estimates for S are easy to obtain but it was Lieb (1967*a*) who first showed rigorously that

$$\lim_{N\to\infty} Z_0^{1/N} = (4/3)^{3/2} \tag{6.6.2}$$

for a square lattice with N vertices. As yet there is no really simple derivation of this result.

An obvious generalization of the six-vertex model is to allow all possible $2^4 = 16$ arrow configurations per vertex on the square lattice. Assigning energies ε_i, $i = 1, 2, \ldots, 16$ to these possible configurations, the partition function for the general 16-*vertex model* can be written as

$$Z = \sum_{\mathrm{C}} \exp\left(-\beta \sum_{i=1}^{16} n_i \varepsilon_i\right) \tag{6.6.3}$$

where n_i is the number of vertices of type i in a given configuration C and the first sum is over all possible arrow configurations on the lattice.

The limiting case $\varepsilon_i \to \infty$, or equivalently, vertex i having *weight*

$$w_i = \exp(-\beta \varepsilon_i) \to 0 \tag{6.6.4}$$

means that the vertex i in question is excluded from the sum over configurations. For example, if all vertex weights except for the six vertices shown in Fig. 6.2 are set equal to zero, we have a class of models known generally as *ferroelectric models with the ice rule.* In particular, if the six energies ε_i, $i = 1, 2, \ldots, 6$ associated with the vertices in Fig. 6.2 are then set equal to zero we have the original ice problem. The F model and the KDP model on a square lattice, which were also solved exactly by Lieb (1967*b*, *c*) have energy assignments given in Table 6.1.

Table 6.1 Ice-rule vertex energies

	ε_1	ε_2	ε_3	ε_4	ε_5	ε_6	$\varepsilon_j, j>6$
Ice	0	0	0	0	0	0	∞
F model	e	e	e	e	0	0	∞
KDP model	0	0	e	e	e	e	∞

If we view the arrows on the bonds as representing a 'polarization' it will be seen, from Fig. 6.2, that vertices 1 to 4 have a net polarization whereas vertices 5 and 6 have zero polarization. The KDP model favours vertices 1 and 2 (with larger weight) and hence has a completely polarized ferroelectric ground state. The F model on the other hand favours vertices 5 and 6 and has an antiferroelectric ground state. The KDP model is in fact a two-dimensional version of a model proposed by Slater (1941) for KH_2PO_4 and the F model proposed by Rys (1963) is a model for two-dimensional antiferroelectrics.

Related models have been considered and solved. Excellent reviews of the models can be found in Lieb and Wu (1972) and Baxter (1982). We will not enter into details here but will discuss the nature of the phase transitions for the F and KDP models, in particular, in the following section.

Another important special case of the 16-vertex model is the *eight-vertex model* which has non-zero weights only for the six ice-rule vertices and the additional two vertices with all four arrows either pointing towards or away from the vertex, as shown in Fig. 6.3.

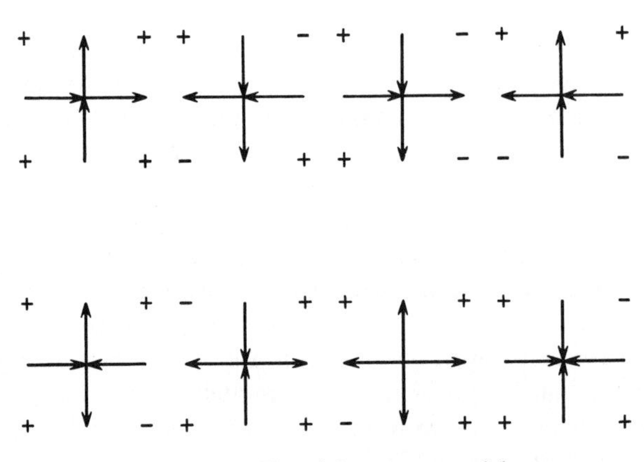

FIG. 6.3 The eight-vertex model.

The plus and minus signs, also shown in Fig. 6.3, are in anticipation of the next step where we show that the eight-vertex model is equivalent to an Ising model with nearest- and next-nearest-neighbour pair spin interactions and a four-spin interaction.

The rule for assigning 'spin up' $(+)$ or 'spin down' $(-)$ to a quadrant or face in Fig. 6.3 is that an arrow to the right or up separates a pair of like spins and an arrow to the left or down separates a pair of unlike spins. Notice that with this rule there are two possible spin configurations per vertex with one obtained from the other by reversing the four-spins surrounding a vertex. (Only one such configuration for each vertex is shown in Fig. 6.3.)

In order to obtain an Ising model which is equivalent to the eight-vertex model, we first consider the spins in Fig. 6.3 to be located on the vertices of the 'dual lattice' with dotted bonds as shown in Fig. 6.4.

Consider now the interaction structure per cell of the dual lattice as shown in Fig. 6.5 with the dotted curve representing a four-spin interaction

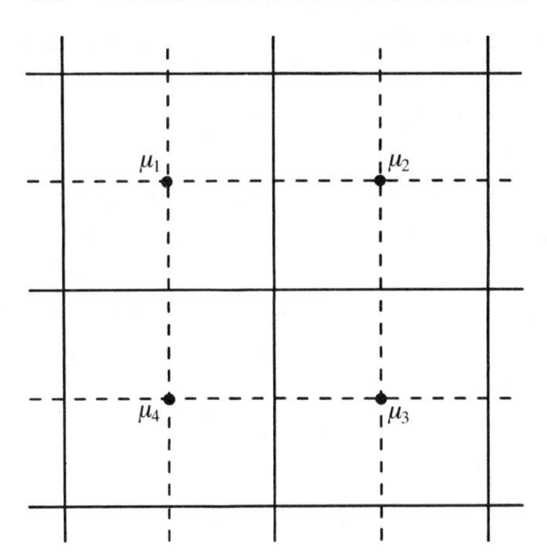

FIG. 6.4 The dual lattice.

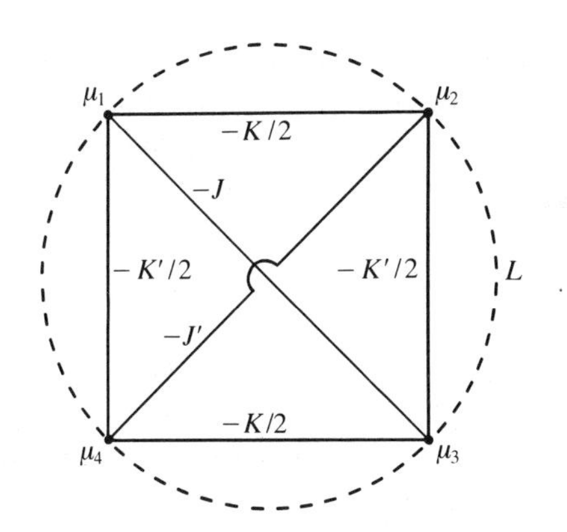

FIG. 6.5 Interaction structure per cell of the dual lattice.

$-L\mu_1\mu_2\mu_3\mu_4$. That is

$$e(\mu_1, \mu_2, \mu_3, \mu_4) = -e_0 - \tfrac{1}{2}K(\mu_1\mu_2 + \mu_3\mu_4) - \tfrac{1}{2}K'(\mu_2\mu_3 + \mu_4\mu_1)$$
$$-J\mu_1\mu_3 - J'\mu_2\mu_4 - L\mu_1\mu_2\mu_3\mu_4 \tag{6.6.5}$$

is the interaction energy per cell (or 'interaction round a face') where e_0 is a constant. The total interaction energy of the Ising model so constructed is then the sum of eqn (6.6.5) over all N cells of the lattice. Noting that nearest-neighbour bonds ($\mu_1\mu_2$, $\mu_2\mu_3$, $\mu_3\mu_4$, and $\mu_4\mu_1$) are shared between neighbouring cells, the Ising model defined by eqn (6.6.5) actually has nearest-neighbour horizontal coupling constant $-K$, nearest-neighbour vertical coupling constant $-K'$, next-nearest-neighbour coupling constants $-J$ and $-J'$, and a four-spin coupling constant $-L$.

Comparing Figs. 6.3 and 6.5, and noting eqn (6.6.5), the identification with the eight-vertex model is obtained by equating vertex energy e_1 with $e(+,+,+,+)$ (or $e(-,-,-,-)$), e_2 with $e(+,-,+,-)$ (or $e(-,+,-,+)$), and so forth. That is

$$\left.\begin{aligned} e_1 &= -e_0 - K - K' - J - J' - L \\ e_2 &= -e_0 + K + K' - J - J' - L \\ e_3 &= -e_0 + K - K' + J + J' - L \\ e_4 &= -e_0 - K + K' + J + J' - L \\ e_5 &= e_6 = -e_0 + J' - J + L \\ e_7 &= e_8 = -e_0 - J' + J + L \end{aligned}\right\} \tag{6.6.6}$$

(Note that in a given configuration on the lattice the number of vertices 5 and 6 are equal as are 7 and 8, so there is no loss of generality in taking $e_5 = e_6$ and $e_7 = e_8$.)

Finally, eqns (6.6.6) can be solved for e_0, K, K', J, J', and L in terms of e_1, e_2, e_3, e_4, e_6, and e_8 and, in view of the twofold degeneracy of spin vs vertex configurations, the partition function Z for the eight-vertex model can be written as

$$Z = 2Z_{\mathrm{I}}, \tag{6.6.7}$$

where Z_{I} is the partition function of the Ising model with cell interaction eqn (6.6.5).

The eight-vertex model of course includes the ferroelectric six-vertex models as special cases and also, with $L = 0$, the two-dimensional Ising model with nearest- and next-nearest-neighbour interactions. Notice that it does not include the Ising model in an external magnetic field.

The general eight-vertex problem has not been solved exactly but Baxter's exact solution (Baxter 1971, 1972) for the special case

$$\begin{gathered} e_1 = e_2,\ e(+,+,+,+)\ e_3 = e_4,\ e(+,+,+,+) \\ e_5 = e_6,\ e(+,+,+,+) \text{ and } e(+,+,+,+)\ e_7 = e_8 \end{gathered} \tag{6.6.8}$$

ranks as an outstanding achievement.

Before discussing Baxter's solution it is worth noting that $e_1 = e_2$ and $e_3 = e_4$ imply $K = K' = 0$. In this case, when $L = 0$, it will be seen from Fig. 6.5 that the model decouples into two independent nearest-neighbour Ising models with coupling constants $-J$ and $-J'$, respectively. The *remarkable thing about the Baxter solution is that when $L \neq 0$ it yields critical exponents that depend on the coupling constants.* This came as a complete shock since it was commonly believed that critical exponents, if not pure rational numbers, should at least be independent of the strength of the microscopic interactions.

6.7 Baxter's solution of the eight-vertex model

The details of Baxter's solution are much too complicated to present here. The basic idea, however, is to set the model up as a transfer matrix problem in much the same way as we did in Section 6.4 for the two-dimensional Ising model. Readers interested in the technical details are referred to Baxter (1982).

Consider a particular column of m horizontal row-bonds in the lattice and let $\sigma_J = +$ or $-$, depending on whether there is an arrow to the right or to the left, respectively, in row J. $\sigma = (\sigma_1, \sigma_2, \ldots, \sigma_m)$ denotes a column configuration of horizontal bonds and has 2^m possible values.

Suppose now that σ and σ' are configurations of two successive columns of bonds. Define the $2^m \times 2^m$ *transfer matrix* $\mathbb{T}$ with elements

$$T(\sigma, \sigma') = \sum_{\mathrm{C}} \exp\left(-\beta \sum_{j=1}^{8} n_j \varepsilon_j\right), \tag{6.7.1}$$

where the sum is over allowed arrangements of arrows on the intervening column of vertical bonds that are consistent with the prescribed arrows on the connecting horizontal bonds specified by σ and σ', and n_j is the number of vertices of type j in this column. Then, if there are n columns with the last column coupled to the first, it is clear that the partition function of the eight-vertex model on an $m \times n$ lattice can be written as

$$\begin{aligned} Z_{m,n} &= \sum_{\sigma_1} \ldots \sum_{\sigma_n} T(\sigma_1, \sigma_2) T(\sigma_2, \sigma_3) \ldots T(\sigma_n, \sigma_1) \\ &= \mathrm{tr}(\mathbb{T}^n) \end{aligned} \tag{6.7.2}$$

where σ_j represents the arrow configuration of the jth column of horizontal bonds.

It follows immediately from the discussion in Section 6.2 and eqn (6.4.6) that the free energy per vertex ψ in the thermodynamic limit is given by

$$-\beta\psi = \lim_{m,n \to \infty} (nm)^{-1} \ln Z_{m,n} = \lim_{m \to \infty} m^{-1} \ln \lambda_1 \tag{6.7.3}$$

where λ_1 is the maximum eigenvalue of the transfer matrix eqn (6.7.1).

Baxter's great achievement was to actually evaluate λ_1. His final result is rather complicated but well worth recording. First we define

$$w_1 = \tfrac{1}{2}(c+d), \quad w_2 = \tfrac{1}{2}(c-d), \quad w_3 = \tfrac{1}{2}(a+b), \quad w_4 = \tfrac{1}{2}(a-b) \tag{6.7.4}$$

where

$$a = \mathrm{e}^{-\beta\varepsilon_1} = \mathrm{e}^{-\beta\varepsilon_2}, \quad b = \mathrm{e}^{-\beta\varepsilon_3} = \mathrm{e}^{-\beta\varepsilon_4}, \quad c = \mathrm{e}^{-\beta\varepsilon_5} = \mathrm{e}^{-\beta\varepsilon_6},$$

and

$$d = \mathrm{e}^{-\beta\varepsilon_7} = \mathrm{e}^{-\beta\varepsilon_8} \tag{6.7.5}$$

are the vertex weights.

In view of symmetry properties of the partition function nothing is changed by replacing each w_i by its absolute value and rearranging and labelling them so that

$$w_1 \geqslant w_2 \geqslant w_3 \geqslant w_4 \geqslant 0. \tag{6.7.6}$$

We now define parameters l, ζ, and V by

$$l=[(w_1^2-w_4^2)(w_2^2-w_3^2)/(w_1^2-w_3^2)(w_2^2-w_4^2)]^{1/2}, \tag{6.7.7}$$

$$\mathrm{sn}(\zeta, l)=[(w_1^2-w_3^2)/(w_1^2-w_4^2)]^{1/2}, \tag{6.7.8}$$

and

$$\mathrm{sn}(v, l)=w_3^{-1}w_4\,\mathrm{sn}(\zeta, l) \tag{6.7.9}$$

where $\mathrm{sn}(u, l)$ is a Jacobi elliptic function of modulus l.

With the ordering eqn (6.7.6), l, ζ, and v are real and satisfy

$$0<l<1 \qquad \text{and} \qquad |v|<\zeta<K(l) \tag{6.7.10}$$

where $K(l)$ is the complete elliptic integral of the first kind defined previously with modulus l.

The final result for the free energy ψ is

$$-\beta\psi=\ln(w_1+w_2)+2\sum_{n=1}^{\infty}\frac{\sinh^2[(\tau-\lambda)n]\,\{\cosh(n\lambda)-\cosh(n\alpha)\}}{n\sinh(2n\tau)\cosh(n\lambda)} \tag{6.7.11}$$

where

$$\tau=\pi K(l)/K(l'), \qquad \lambda=\pi\zeta/K(l'), \qquad \alpha=\pi v/K(l') \tag{6.7.12}$$

and $l'=(1-l^2)^{1/2}$ is the complementary modulus to l.

It is more or less clear from eqns (6.7.7)–(6.7.9) that a phase transition in this model will occur when two or more w_js become equal. In general, for arbitrary weights a, b, c, and d, only two ws can become numerically equal by varying the temperature. When three are equal or two pairs are equal, the situation is more complicated and this is indeed what happens with the F and KDP models where $d=0$. We will discuss these two models later.

In general, when $w_1=w_2$ or $w_3=w_4$, the free energy is analytic. When the middle two ws are equal, however, we have a phase transition at a critical temperature T_c, with the singular part of the free energy ψ_s having the form

$$\psi_s \sim \cot(\tfrac{1}{2}\pi^2/\bar{\mu})|T-T_c|^{\pi/\bar{\mu}} \qquad \text{as } T\to T_c \tag{6.7.13}$$

or, if $\pi/2\bar{\mu}=m$, an integer,

$$\psi_s \sim \pi^{-1}2(T-T_c)^{2m}\ln|T-T_c| \qquad \text{as } T\to T_c \tag{6.7.14}$$

where $(0<\bar{\mu}<\pi)$,

$$\bar{\mu}=\pi\zeta/K(l) \qquad \text{evaluated at } T=T_c. \tag{6.7.15}$$

From eqn (6.7.7), $T=T_c$ (or $w_2=w_3$) corresponds to $l=0$, so $K(l)=\pi/2$,

$\mathrm{sn}(u, l) = \sin u$ and from eqns (6.7.8), (6.7.15), and (6.7.4),

$$\cos \bar{\mu} = (2w_2^2 - w_1^2 - w_4^2)/(w_1^2 - w_4^2) = (ab - cd)/(ab + cd). \qquad (6.7.16)$$

Another exceptional case occurs if $\pi/\bar{\mu}$ is an odd integer greater than unity in which case all singularities in the free energy disappear. In general, however, from eqns (6.7.13) and (6.7.16), the free energy has a branch-point singularity with an exponent $\pi/\bar{\mu}$ that can be varied continuously from one to infinity by an appropriate choice of interaction parameters.

In terms of the equivalent spin model with nearest-neighbour coupling constants J, J' (>0) and four-spin coupling constant $L(>0)$ the parameters w_j arranged in decreasing order, from eqns (6.6.6), (6.7.4), and (6.7.5) are

$$w_1 = A\mathrm{e}^{\beta L} \cosh \beta(J + J'), \qquad w_2 = A\mathrm{e}^{\beta L} \sinh \beta(J + J'),$$

$$w_3 = A\mathrm{e}^{-\beta L} \cosh \beta(J - J'), \quad w_4 = A\mathrm{e}^{-\beta L} \sinh \beta|J - J'| \qquad (6.7.17)$$

(with perhaps w_2 and w_3 interchanged). The critical point ($w_2 = w_3$) corresponds to

$$\sinh \beta(J + J') = \mathrm{e}^{-2\beta L} \cosh \beta(J - J') \qquad (6.7.18)$$

and the critical exponent parameter $\bar{\mu}$ from eqn (6.7.16) is given by

$$\cos \bar{\mu} = \tanh(2\beta L) \qquad (6.7.19)$$

so that the critical exponents are primarily dependent on the four-spin coupling constant L. In particular, when $L = 0$, $\bar{\mu} = \pi/2$, so that from eqn (6.7.14),

$$\psi_{\mathrm{s}} \sim \pi^{-1} 2(T - T_{\mathrm{c}})^2 \ln|T - T_{\mathrm{c}}|, \qquad (6.7.20)$$

which agrees, of course, as it should with Onsager.

With the results so far, we have the critical exponents α and α', for the high- and low-temperature specific heats, given from eqn (6.7.13) by

$$\alpha = \alpha' = 2 - \pi/\bar{\mu} \qquad (6.7.21)$$

or, in the case of Onsager, eqn (6.7.20), a logarithmic divergence.

The correlation-length exponent for the eight-vertex model is also known exactly (Baxter 1973) and takes the value

$$\nu = \pi/2\bar{\mu}. \qquad (6.7.22)$$

On the basis of series expansions and in the belief that natural variables for two-dimensional models are given in terms of elliptic functions or integrals as above, Barber and Baxter (1973) were led naturally to conjecture an analytic expression for the spontaneous magnetization (or polarization) yielding the critical exponent

$$\beta = \pi/16\bar{\mu}. \qquad (6.7.23)$$

Assuming, by analogy with the two-dimensional Ising model, that the exponent relations

$$\alpha+2\beta+\gamma=2, \qquad \alpha+\beta(1+\delta)=2, \qquad \text{and} \qquad \gamma=(2-\eta)\nu, \quad (6.7.24)$$

are valid, we deduce from eqns (6.7.21) and (6.7.23) (Barber and Baxter 1973) that

$$\gamma=7\pi/8\bar{\mu}, \qquad \delta=15, \qquad \text{and} \qquad \eta=1/4. \quad (6.7.25)$$

The results (6.7.23) and (6.7.25) are exact, of course, for the nearest-neighbour Ising model ($\bar{\mu}=\pi/2$ giving $\beta=1/8$ and $\gamma=7/4$) but so far questionable for the eight-vertex model.

It is interesting that the coupling-dependent parameter $\bar{\mu}$ appears in the exact and conjectured exponents above in the same way and that the exponents δ and η, which are defined *at* the critical temperature, are independent of $\bar{\mu}$. This is very intriguing, and suggestive of what critical exponents may ultimately depend upon.

Returning finally to various special cases of the eight-vertex model, we notice in particular that, for the F model and the KDP model, the vertex weight d, given by eqn (6.7.5), is zero so that, from eqn (6.7.4), the parameters w_1 and w_2 are identical. In terms of critical criteria for the eight-vertex model this means that we are already in the realm of pathology so far as the discussion above for the eight-vertex model is concerned.

Notice, in particular, that when $d \to 0$, $\bar{\mu}$, from eqn (6.7.16), approaches zero or π which correspond to the extremes of the range available to $\bar{\mu}$ for the eight-vertex model.

One could view the limiting value $\bar{\mu}=0$ as representing a phase transition of infinite order, which is in fact the case for the F model. Moreover, the limiting free energy for the F model has a branch point at a finite critical temperature T_c with the cut a natural boundary. The phase transition is of infinite order in the sense that the free energy and all its derivatives are finite and continuous at T_c. The expansion of the free energy around T_c, however, has zero radius of convergence.

The phase transition for the KDP model is also rather peculiar. The specific heat diverges as $(T-T_c)^{-1/2}$, when $T \to T_c+$ and is identically zero for $T<T_c$. The model also has a latent heat (i.e. a discontinuity in the internal energy) corresponding to a first-order phase transition.

6.8 Hard-core lattice systems

A hard-core lattice system is a special case of a lattice gas, defined in general in Section 3.4, in which there is infinite repulsion between neighbouring sites. In other words, if a lattice site is occupied, none of its nearest-neighbour sites can be occupied. An example is shown in Fig. 6.6 for the special case of a square lattice. Since the occupation of a particular site (circled) excludes the

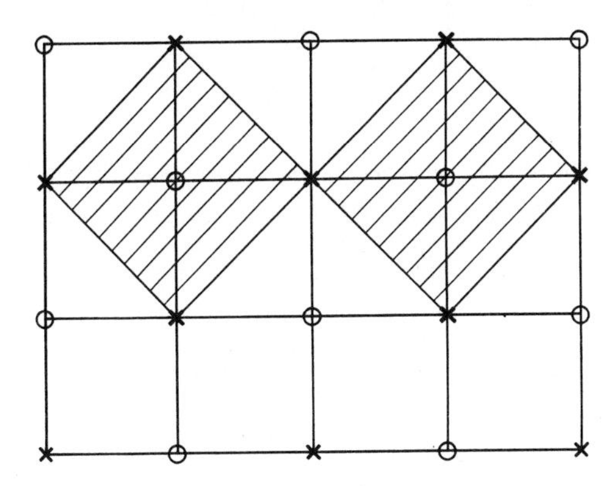

FIG. 6.6 The hard-square gas.

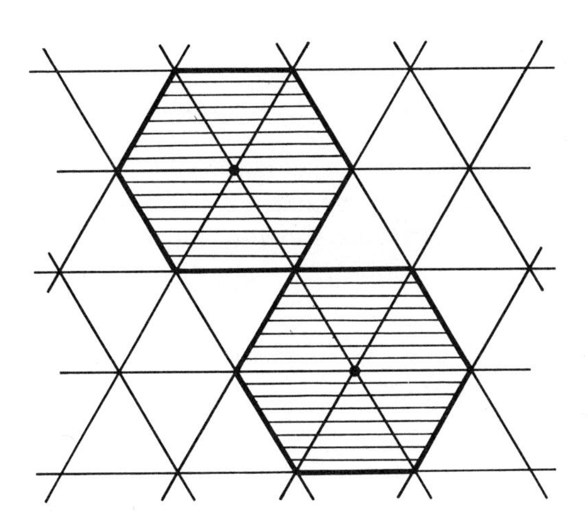

FIG. 6.7 Hard hexagons on a triangular lattice.

occupation of the four neighbouring sites (crossed), we can think of this as a gas of *hard squares* shown as shaded figures in Fig. 6.6. Similarly, the hard-core system on a triangular lattice as shown in Fig. 6.7 may be thought of as a gas of *hard hexagons.*

In general, the grand-canonical partition function for a lattice gas of V sites is given by

$$Z_G(V, T, z) = \sum_{N=0}^{\infty} z^N \sum_{\{t\}}{}' \exp(-\beta E\{t\}) \tag{6.8.1}$$

where the primed sum over configurations $\{t\} = (t_1, t_2, \ldots, t_V)$ is subject to the condition

$$\sum_{i=1}^{V} t_i = N \tag{6.8.2}$$

which restricts the total number of occupied sites to be N. $E\{t\}$ in eqn (6.8.1) is the interaction energy and the configuration variables t_i are defined by

$$t_i = \begin{cases} 1 & \text{if the } i\text{th site is occupied} \\ 0 & \text{if the } i\text{th site is unoccupied.} \end{cases} \tag{6.8.3}$$

In the case of infinite repulsion between neighbouring sites, the Boltzmann factor

$$\exp(-\beta E\{\mu\}) = \begin{cases} 0 & \text{if } t_i t_j = 1 \text{ for any two nearest-neighbour sites } i \text{ and } j \\ 1 & \text{otherwise.} \end{cases} \tag{6.8.4}$$

It thus follows from eqn (6.8.1) that

$$Z_G(V, T, z) = \sum_{N=0}^{\infty} z^N G(V, N) \tag{6.8.5}$$

where $G(V, N)$ is the number of ways of placing N particles on V sites such that no two particles occupy the same site or any two nearest-neighbour sites.

Alternatively, if we use eqn (6.8.2) to write eqn (6.8.1) in the form

$$Z_G(V, T, z) = \sum_{\{t\}} z^{t_1 + t_2 + \ldots + t_V} \exp(-\beta E\{t\}) \tag{6.8.6}$$

where the sum over $\{t\}$ is now over all possible 2^V configurations, the hard-core system from eqn (6.8.4) has

$$Z_G(V, z) = \sum_{\{t\}} z^{t_1 + t_2 + \ldots + t_V} \prod_{(ij)} (1 - t_i t_j) \tag{6.8.7}$$

where the product over (ij) denotes a product over nearest-neighbour lattice sites i and j.

Written in the form of eqn (6.8.7) it is clear that the transfer-matrix method can be applied to hard-core lattice systems in general. In particular, for the square lattice with m rows and n columns, the grand-canonical potential is given by (with $V = nm$)

$$\chi(z) = \lim_{V \to \infty} V^{-1} \ln Z_G(V, z) = \lim_{m \to \infty} m^{-1} \ln \lambda_1 \tag{6.8.8}$$

where λ_1 is the maximum eigenvalue of the column-to-column transfer matrix $\mathbb{T}$ with elements

$$T(t, t') = \prod_{i=1}^{m-1} (1 - t_i t_{i+1})(1 - t'_i t'_{i+1}) \prod_{i=1}^{m} z^{(t_i + t'_i)/2} (1 - t_i t'_i) \tag{6.8.9}$$

where $t=(t_1, t_2, \ldots, t_m)$ and $t'=(t'_1, t'_2, \ldots, t'_m)$ denote configurations of nearest-neighbour columns.

It will be noted that $\chi(z)$ for hard-core systems is independent of temperature and depends only on the fugacity z.

The density ρ in the grand-canonical description is given in terms of z by

$$\rho = z\frac{\partial \chi}{\partial z} \tag{6.8.10}$$

and the pressure by

$$p = kT\chi. \tag{6.8.11}$$

The pressure as a function of density is then obtained by eliminating z between eqns (6.8.10) and (6.8.11).

If a hard-core system undergoes a phase transition, it follows that p must have a singular point at some critical density ρ_c less than the close-packing density or, equivalently, that $\chi(z)$ is singular at some finite positive value z_c. The transition is then interpreted as a gas–solid transition.

The order parameter, analogous to m_0 for a magnet or $\rho_\ell - \rho_g$ for a fluid, is specified by noting that there are usually sublattice structures corresponding to distinct close-packing configurations on the lattice. For the square lattice shown in Fig. 6.6, for example, there are two close-packed configurations corresponding to the particles occupying sites marked as circles or as crosses. Clearly, if one imposes the condition that all sites of one sublattice on the boundary be occupied, then that sublattice will form the close-packed configuration. For densities smaller than the close-packing density but larger than ρ_c or, equivalently, for $z > z_c$, one would expect that with such a boundary condition the difference in mean densities of the two sublattices would be non-zero while for $\rho < \rho_c$ or $z < z_c$ there would be no difference. Specifically, if site 1 is on the preferred sublattice and site 2 on the other sublattice, one would expect

$$\rho_1 - \rho_2 > 0 \qquad \text{for } z > z_c \tag{6.8.12}$$

where ρ_1 and ρ_2 are the respective mean sublattice densities per site defined by

$$\begin{aligned}\rho_k &= \langle t_k \rangle \\ &= \lim_{V\to\infty} Z_G^{-1} \sum_{\{t\}} t_k z^{t_1 + \cdots + t_V} \prod_{(ij)} (1 - t_i t_j), \qquad k = 1, 2.\end{aligned} \tag{6.8.13}$$

Moreover, one might expect

$$\rho_1 = \rho_2 \qquad \text{for } z < z_c \tag{6.8.14}$$

and

$$\rho_1 - \rho_2 \sim A(z - z_c)^\beta \qquad \text{as } z \to z_c+. \tag{6.8.15}$$

This defines the order parameter exponent β in complete analogy with the magnet and fluid transition, with the fugacity z replacing the temperature as the critical variable.

Similarly, the exponents α and α', which characterize the critical behaviour of the second derivative of the free energy with respect to T (i.e. the specific heat) for magnets and fluids, are now defined in terms of the second derivative of $\chi(z)$ with respect to z. That is

$$\frac{d^2\chi}{dz^2} \sim \begin{cases} c_-\left|1-\dfrac{z}{z_c}\right|^{-\alpha} & \text{as } z \to z_c- \\ c_+\left|1-\dfrac{z}{z_c}\right|^{-\alpha'} & \text{as } z \to z_c+ \end{cases} \tag{6.8.16}$$

and, similarly, for the correlation-length exponent ν.

It turns out that the hard-square lattice gas has not yet been solved exactly. Baxter (1980), however, obtained the exact solution of the hard-hexagon model. Since a detailed account of the derivation and results is given elsewhere (Baxter 1982) we will not enter into any detail except to record the values of the critical exponents defined above,

$$\alpha = \alpha' = 1/3, \qquad \beta = 1/9, \qquad \text{and} \qquad \nu = 5/6. \tag{6.8.17}$$

We also note that for the hexagonal lattice there are, in fact, three inequivalent sublattices so that, in many respects, the hard-hexagon model is more complicated than the hard-square model.

6.9 The Gaussian and spherical models

The only non-trivial statistical mechanical models that have been solved exactly in dimensions greater than two are the so-called Gaussian and spherical models which were invented by Kac in 1947 (see Kac 1964).

Kac's idea was in fact to simplify the Ising problem to the point where it could be solved exactly in any dimension. His first simplification consisted of replacing the Ising spins by Gaussian random variables. Thus consider a set of 'spins' $x_{\mathbf{r}}$ located on the vertices $\mathbf{r}=(r_1, r_2, \ldots, r_d)$ of a regular d-dimensional hypercubic lattice of $N=n^d$ sites and with interaction energy

$$E\{x\} = -\sum_{\mathbf{r}\neq\mathbf{r}'} J_{\mathbf{r},\mathbf{r}'} x_{\mathbf{r}} x_{\mathbf{r}'} - H\sum_{\mathbf{r}} x_{\mathbf{r}} \tag{6.9.1}$$

where the sums extend over all lattice vectors

$$\mathbf{r} = (r_1, r_2, \ldots, r_d); \qquad r_i = 1, 2, \ldots, n, \quad i = 1, 2, \ldots, d. \tag{6.9.2}$$

Assuming that the spins $x_{\mathbf{r}}$ are independent Gaussian random variables with arbitrary variance, the partition function for this model, called obviously the Gaussian model, is given by

$$Z_N(\beta, H) = \int_{-\infty}^{\infty}\!\!\cdots\!\int \exp(-\beta E\{x\}) \prod_{\mathbf{r}} \exp(-s x_{\mathbf{r}}^2) \mathrm{d}x_{\mathbf{r}}. \tag{6.9.3}$$

In order to evaluate the integral in eqn (6.9.3) we assume, for simplicity, that the system is translationally invariant and periodic in the d coordinate directions. That is, the coupling constants $J_{\mathbf{r},\mathbf{r}'}$ in eqn (6.9.1) have the form

$$J_{\mathbf{r},\mathbf{r}'}=J(\mathbf{r}-\mathbf{r}') \tag{6.9.4}$$

and satisfy

$$J(\mathbf{r}+\mathbf{n})=J(\mathbf{r})=J(-\mathbf{r}) \tag{6.9.5}$$

where

$$\mathbf{n}=(n, n, \ldots, n). \tag{6.9.6}$$

The coupling constants $J_{\mathbf{r},\mathbf{r}'}$ then form a generalized cyclic matrix which can be diagonalized by a Fourier transform. Specifically, for $J_{\mathbf{r},\mathbf{r}'}$ satisfying eqns (6.9.4)–(6.9.6),

$$(S^{-1}JS)_{\mathbf{k},\ell}=\delta_{\mathbf{k},\ell}\,\lambda(2\pi\mathbf{k}/n) \tag{6.9.7}$$

where $\mathbf{k}$ and ℓ are vectors of the form given in eqn (6.9.2),

$$S_{\mathbf{k},\ell}=n^{-d/2}\prod_{\alpha=1}^{d}\exp(2\pi i k_\alpha l_\alpha/n) \tag{6.9.8}$$

and the eigenvalues λ of the matrix $(J_{\mathbf{r},\mathbf{r}'})$, given by

$$\lambda(\boldsymbol{\phi})=\sum_{\mathbf{r}}J(\mathbf{r})\exp(i\mathbf{r}\cdot\boldsymbol{\phi}), \tag{6.9.9}$$

are just the generalized Fourier coefficients of $J_{\mathbf{r}}$.

After an orthogonal change of variables in the integral eqn (6.9.3), which diagonalizes the exponent of the integrand, we are left with the problem of evaluating the multiple Gaussian integral

$$Z_N(\beta, H)=\int_{-\infty}^{\infty}\cdots\int\exp\left\{-\sum_{\mathbf{k}}[s-\beta\lambda(2\pi\mathbf{k}/n)]z_{\mathbf{k}}^2+N^{1/2}\beta H z_{\mathbf{0}}\right\}\prod_{\mathbf{k}}dz_{\mathbf{k}}. \tag{6.9.10}$$

Since the integrand in eqn (6.9.10) factors, we only need the elementary result

$$\int_{-\infty}^{\infty}\exp(-ax^2+bx)dx=(\pi/a)^{1/2}\exp(b^2/4a), \tag{6.9.11}$$

which holds for any b and $a>0$, to obtain

$$Z_N(\beta, H)=\prod_{\mathbf{k}}\{\pi/[s-\beta\lambda(2\pi\mathbf{k}/n)]\}^{1/2}\exp\{N(\beta H)^2/4[s-\beta\lambda(\mathbf{0})]\} \tag{6.9.12}$$

provided that

$$s>\beta\max_{\boldsymbol{\phi}}\lambda(\boldsymbol{\phi}). \tag{6.9.13}$$

In the thermodynamic limit, $N=n^d\to\infty$, the free energy per spin $\psi(\beta, H)$ is then given by

$$-\beta\psi(\beta, H)=\lim_{N\to\infty} N^{-1}\ln Z_N(\beta, H)$$

$$=-\lim_{n\to\infty}(2n^d)^{-1}\sum_{k_1=1}^{n}\cdots\sum_{k_d=1}^{n}\ln\left[s-\beta\lambda\left(\frac{2\pi k_1}{n},\ldots,\frac{2\pi k_d}{n}\right)\right]$$

$$+(\beta H)^2/4[s-\beta\lambda(\mathbf{0})]+\frac{1}{2}\ln\pi \tag{6.9.14}$$

$$=-\frac{1}{2}(2\pi)^{-d}\int_0^{2\pi}\!\!\cdots\!\int\ln(s-\beta\lambda(\boldsymbol{\phi}))\mathrm{d}\phi_1\cdots\mathrm{d}\phi_d$$

$$+(\beta H)^2/4[s-\beta\lambda(\mathbf{0})]+\frac{1}{2}\ln\pi \tag{6.9.15}$$

where $\lambda(\boldsymbol{\phi})$ is defined by eqn (6.9.9) with the sum now taken over the infinite lattice.

As an example, if we have nearest-neighbour interactions only, $J_{\mathbf{r},\mathbf{r}'}=J/2$ when $|\mathbf{r}-\mathbf{r}'|=1$ and zero otherwise. That is, for $\mathbf{r}=(r_1,\ldots,r_d)$

$$J(\mathbf{r})=\begin{cases}J/2 & \text{when } r_i=\pm 1, r_j=0, \quad 1\leqslant i\neq j\leqslant d\\ 0 & \text{otherwise.}\end{cases} \tag{6.9.16}$$

In this case we have from eqn (6.9.9) that

$$\lambda(\phi_1,\phi_2,\ldots,\phi_d)=J\sum_{i=1}^{d}\cos\phi_i \tag{6.9.17}$$

and hence, from eqn (6.9.13), that the free energy is given by eqn (6.9.15) provided

$$s>d\beta J. \tag{6.9.18}$$

Kac called this the low-temperature catastrophe since for any given $s>0$, the free energy of the Gaussian model with nearest-neighbour interactions is not defined for temperatures smaller than the critical value $T_c=dJ/ks$ corresponding to equality in eqn (6.9.18). Obviously, any Gaussian model, for which the maximum (or supremum) of $\lambda(\boldsymbol{\phi})$ is positive, will suffer from the same catastrophe.

In order to overcome this difficulty and to construct a model which was more Ising-like, Kac next allowed the spins in the interaction-energy eqn (6.9.1) to take on any real value but subject to the overall *spherical constraint*

$$\sum_{\mathbf{r}} x_{\mathbf{r}}^2=N \tag{6.9.19}$$

where N is the total number of spins. It is to be noted that Ising spins ($x_{\mathbf{r}}=\pm 1$) automatically satisfy the constraint (6.9.19) but, unlike the so-called spherical model whose configurations are distributed uniformly on the sphere (6.9.19), the corresponding Ising model has configurations restricted to the 2^N vertices $(\pm 1, \pm 1, \ldots, \pm 1)$ of the N-dimensional hypercube inscribed in the sphere (6.9.19). The partition function for the spherical model is thus given by

$$Z_N(\beta, H) = \int \exp(-\beta E\{x\})\,\mathrm{d}\sigma_{\sqrt{N}} \tag{6.9.20}$$

where the integral is taken over the surface of the sphere (6.9.19).

In a masterful piece of asymptotic analysis, Berlin and Kac (1952) managed to obtain the limiting free energy of the spherical model for nearest-neighbour interactions in one, two, and three dimensions. It was clear, however, that their method could be applied to the spherical model with more general interactions in arbitrary dimensions.

Shortly after the publication of the Berlin and Kac work, Lewis and Wannier (1952) pointed out that the same results could be obtained more easily by considering the grand-canonical ensemble in which the spherical constraint (6.9.19) is satisfied 'on the average'. Thus, if we return to eqn (6.9.3) and allow the quantity s to vary, the right-hand side of eqn (6.9.3) may be interpreted as the grand-canonical partition function, redefined as

$$Z_G(N, \beta, H, s) = \int \int_{-\infty}^{\infty} \exp\left(\beta \sum_{\mathbf{r},\mathbf{r}'} J_{\mathbf{r},\mathbf{r}'} x_{\mathbf{r}} x_{\mathbf{r}'} + \beta H \sum_{\mathbf{r}} x_{\mathbf{r}}\right)$$

$$\times \exp\left(-s \sum_{\mathbf{r}} x_{\mathbf{r}}^2\right) \prod_{\mathbf{r}} \mathrm{d}x_{\mathbf{r}}. \tag{6.9.21}$$

Proceeding in the usual way, the spherical constraint (6.9.19) is now satisfied on the average by choosing s in such a way that

$$\left\langle \sum_{\mathbf{r}} x_{\mathbf{r}}^2 \right\rangle \equiv -\frac{\partial}{\partial s} \ln Z_G(N, \beta, H, s) = N. \tag{6.9.22}$$

Although the prescription specified by eqns (6.9.21) and (6.9.22) is still the spherical model, treated in a different ensemble, it is usually referred to as the *mean-spherical model* for obvious reasons. The thermodynamic properties of the two 'models' are identical but there are subtle differences in correlation functions at low temperatures. For details, the interested reader is referred to Joyce (1972) and Kac and Thompson (1977). Here we discuss only the mean-spherical model.

The evaluation of the grand-canonical partition function (6.9.21) has already been carried out in eqn (6.9.12). From this result we obtain

$$N^{-1} \ln Z_G(N, \beta, H, s) = -(2n^d)^{-1} \sum_{k_1=1}^{n} \cdots \sum_{k_d=1}^{n} \ln[s - \beta\lambda(2\pi\mathbf{k}/n)]$$

$$+ (\beta H)^2/4[s - \beta\lambda(\mathbf{0})] + \frac{1}{2}\ln \pi \tag{6.9.23}$$

and hence, from eqn (6.9.22), the parameter s is determined from the condition

$$(2n^d)^{-1} \sum_{k_1=1}^{n} \cdots \sum_{k_d=1}^{n} [s-\beta\lambda(2\pi\mathbf{k}/n)]^{-1}+(\beta H)^2/4[s-\beta\lambda(\mathbf{0})]=1. \quad (6.9.24)$$

For large N we expect $\sum_{\mathbf{r}} x_{\mathbf{r}}^2$ to be sharply peaked about its mean value of N so that, on comparing eqns (6.9.20) and 6.9.21), one has

$$Z_G(N,\beta,H,s) \sim Z_N(\beta,H) \exp(-sN) \qquad \text{as } N \to \infty. \quad (6.9.25)$$

It then follows, from eqns (6.9.23) and (6.9.24) that, in the limit $N \to \infty$, the free energy of the spherical model $\psi(\beta,H)$ is given by

$$-\beta\psi(\beta, H)=s-\frac{1}{2}(2\pi)^{-d} \int_0^{2\pi} \cdots \int \ln[s-\beta\lambda(\boldsymbol{\phi})]\,\mathrm{d}\phi_1 \cdots \mathrm{d}\phi_d$$

$$+(\beta H)^2/4[s-\beta\lambda(\mathbf{0})]+\frac{1}{2}\ln\pi \quad (6.9.26)$$

where, on setting $\zeta = s/\beta$, the quantity s, as a function of β and H, is determined from

$$\beta = R(\zeta)+\beta H^2/4[\zeta-\lambda(\mathbf{0})]^2 \quad (6.9.27)$$

with

$$R(\zeta)=\frac{1}{2}(2\pi)^{-d} \int_0^{2\pi} \cdots \int [\zeta-\lambda(\boldsymbol{\phi})]^{-1}\mathrm{d}\phi_1 \cdots \mathrm{d}\phi_d \quad (6.9.28)$$

and, in order for the above to be valid, we require from eqn (6.9.13) that

$$\zeta > \max \lambda(\boldsymbol{\phi}) \equiv \zeta_c. \quad (6.9.29)$$

The above argument based on eqn (6.9.25) can be made perfectly rigorous by several methods which we will not discuss here (see Joyce 1972; Kac and Thompson 1971). The original derivation of Berlin and Kac used the method of steepest descents and in this context eqn (6.9.27) is usually referred to as the 'saddle-point equation'.

The existence or non-existence of a phase transition for the spherical model depends on the behaviour of the integral $R(\zeta)$, defined by eqn (6.9.28), as $\zeta \to \zeta_c+$. In particular, we will see in a moment that *there is a phase transition in zero field if and only if $R(\zeta_c)$ is finite, with critical point given by*

$$\beta_c = R(\zeta_c). \quad (6.9.30)$$

To see how this comes about consider first the case where $R(\zeta)$ diverges as $\zeta \to \zeta_c+$. Setting the field equal to zero in eqn (6.9.27) and noting that $R(\zeta)$ is monotone decreasing, we see from Fig. 6.8 that for every $\beta > 0$ there is a unique solution $\zeta(\beta) > \zeta_c$ of the saddle-point equation (6.9.27). Moreover, it is

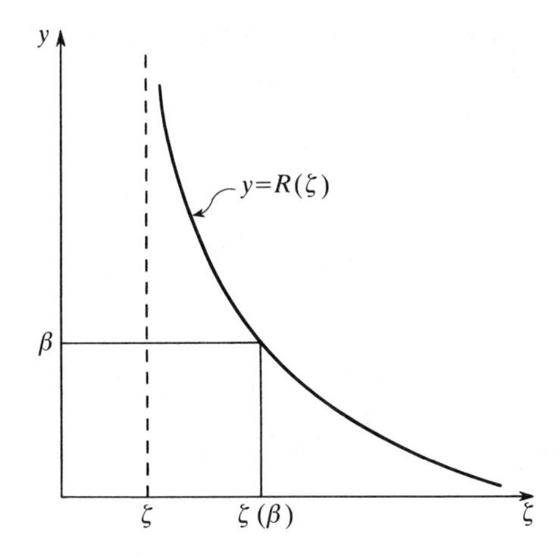

FIG. 6.8 Solution of saddle-point equation when $R(\zeta_c)$ diverges.

clear that in this case $\zeta(\beta)$ and hence $\psi(\beta, 0)$ are analytic functions of β so that there is no phase transition.

When $R(\zeta_c) = \beta_c$ is finite, however, there is a problem with eqn (6.9.27) when $H = 0$ and $\beta > \beta_c$, although intuitively one would 'guess' from Fig. 6.9 that, in this low-temperature region, the saddle point ζ should 'stick' at ζ_c. To see that this is so let us consider the case of ferromagnetic interactions $J(\mathbf{r}) \geqslant 0$ for which

$$\lambda(\boldsymbol{\phi}) = 2\sum_{\mathbf{r}} J(\mathbf{r})\cos(\mathbf{r}\cdot\boldsymbol{\phi}) \leqslant \lambda(\mathbf{0}) \tag{6.9.31}$$

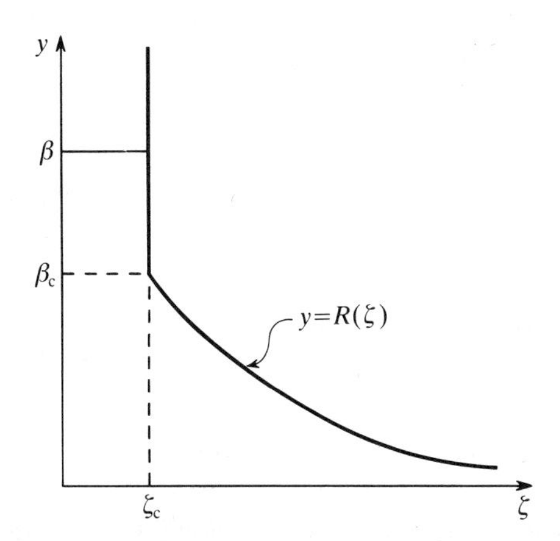

FIG. 6.9 Saddle point sticks at ζ_c when $R(\zeta_c)$ is finite and $\beta > \beta_c$.

and hence from eqn (6.9.29) that $\zeta_c = \lambda(\mathbf{0})$. In this case the right-hand side of the saddle-point equation (6.9.27) diverges as $\zeta \to \zeta_c$ for all $H \neq 0$ and hence, as indicated in Fig. 6.8, the saddle point $\zeta(\beta, H)$ exists and is an analytic function of β and H for all β and $H \neq 0$. As H becomes small, however, it is clear from eqn (6.9.27) that ζ must deviate from $\zeta_c = \lambda(\mathbf{0})$ by something of order H. Thus, if we set

$$\zeta = \lambda(\mathbf{0}) + H/2\eta \qquad (\eta > 0, H > 0) \tag{6.9.32}$$

in eqn (6.9.27), we obtain

$$\beta = R(\zeta_c + H/2\eta) + \beta\eta^2, \tag{6.9.33}$$

which in the limit $H \to 0+$ gives, from eqn (6.9.30),

$$\eta = \left(1 - \frac{\beta_c}{\beta}\right)^{1/2} \qquad \text{when } \beta > \beta_c, \tag{6.9.34}$$

with the saddle point sticking at ζ_c for low temperatures $T < T_c$.

It will be noted that the above argument only holds when $\zeta_c = \lambda(\mathbf{0})$. When $\zeta_c > \lambda(\mathbf{0})$ some interesting new features appear which result in unusual critical behaviour. The interested reader is referred to Smith and Thompson (1986) for details.

As we will see in the following section, the sticking of the saddle point, which provides the underlying mathematical mechanism for the phase transition in the spherical model, produces unphysical behaviour below the critical point. In fact one could argue that the model itself is unphysical on the grounds that the microscopic variables are subject to a constraint that depends on the size of the system. As pointed out by Stanley (1968), however, the spherical model can be viewed as a limit of a sequence of n-vector models, which have no macroscopic constraint, by fixing the length of each spin to be $n^{1/2}$ and taking the limit $n \to \infty$. Specifically, if one considers the partition function

$$Z_N^{(n)}(\beta, H) = \int_{\{\|\mathbf{S}_\mathbf{r}\| = n^{1/2}\}} \cdots \int \exp\left(\beta \sum_{\mathbf{r},\mathbf{r}'} J_{\mathbf{r},\mathbf{r}'} \mathbf{S}_\mathbf{r} \cdot \mathbf{S}_{\mathbf{r}'} + \beta \mathbf{H} \cdot \sum_\mathbf{r} \mathbf{S}_\mathbf{r}\right) \prod_\mathbf{r} d\mathbf{S}_\mathbf{r} \tag{6.9.35}$$

of the n-vector model where the integrations are performed over the n-dimentional sphere of radius $n^{1/2}$ for each spin separately, then apart from trivial constants,

$$\psi(\beta, H) = -\beta^{-1} \lim_{n,N \to \infty} (nN)^{-1} \ln Z_N^{(n)}(\beta, H) \tag{6.9.36}$$

is the spherical-model free energy, provided the system is translationally invariant. This result was first proved by Kac and Thompson (1971).

6.10 Critical behaviour of the spherical model

Thermodynamic quantities for the spherical model can, as usual, be obtained from the free energy by differentiation with respect to the field and temperature variables. For example, if we differentiate the free energy eqn (6.9.26) with respect to H, noting, in view of the saddle-point condition (6.9.27), that the coefficient of the term $\partial\zeta/\partial H$ vanishes identically, we obtain for the magnetization

$$m(\beta, H) = -\frac{\partial \psi}{\partial H} = H/2[\zeta - \lambda(\mathbf{0})]. \tag{6.10.1}$$

From eqns (6.9.32) and (6.9.34) we then have for the spontaneous magnetization

$$m_0 = \lim_{H\to 0+} m(\beta, H) = \begin{cases} 0 & \text{when } T > T_c \\ \left(1 - \dfrac{T}{T_c}\right)^{1/2} & \text{when } T < T_c. \end{cases} \tag{6.10.2}$$

It follows that whenever a ferromagnetic spherical model has a phase transition the critical exponent β is always equal to the classical value of one-half, regardless of the dimensionality of the system and the form of the potential.

The isothermal susceptibility from eqn (6.10.1) is given by

$$\chi = \frac{\partial m}{\partial H} = [2(\zeta - \lambda(\mathbf{0}))]^{-1} - 2H[2(\zeta - \lambda(\mathbf{0})]^{-1}\frac{\partial \zeta}{\partial H}. \tag{6.10.3}$$

When $T < T_c$, the saddle point sticks at $\zeta = \lambda(\mathbf{0})$ so for temperatures below the critical temperature the zero-field susceptibility is infinite. This is another unusual feature of the spherical model.

When $T > T_c$, the second term in eqn (6.10.3) vanishes in the limit $H \to 0+$; hence the zero-field susceptibility is given by

$$\chi_0 = [2(\zeta - \lambda(\mathbf{0}))]^{-1} \qquad (T > T_c) \tag{6.10.4}$$

where, from eqns (6.9.27) and (6.9.28), $\zeta > \lambda(\mathbf{0})$ is the solution of

$$\beta = R(\zeta) = \tfrac{1}{2}(2\pi)^{-d} \int_0^{2\pi}\!\!\cdots\!\int [\zeta - \lambda(\boldsymbol{\phi})]^{-1}\, \mathrm{d}\phi_1 \cdots \mathrm{d}\phi_d. \tag{6.10.5}$$

Similarly, the zero-field specific heat is given by (Problem 12)

$$C_0 = -T\frac{\partial^2 \psi}{\partial T^2}(\beta, 0) = \begin{cases} k/2 & \text{when } T < T_c \\ \dfrac{k}{2}\left(1 + 2\beta^2 \dfrac{\partial \zeta}{\partial \beta}\right) & \text{when } T > T_c. \end{cases} \tag{6.10.6}$$

The critical behaviour of χ_0 and C_0 as $T \to T_c+$ is now determined by the behaviour of $\zeta = \zeta(\beta)$ as $T \to T_c$ or, equivalently, by the behaviour of the integral $R(\zeta)$ in eqn (6.10.5) as $\zeta \to \lambda(\mathbf{0})+ = \zeta_c+$, which obviously depends on the dimensionality d and the form of the interaction through its Fourier transform $\lambda(\boldsymbol{\phi})$.

In the case of nearest-neighbour interactions only we have seen, from eqns (6.9.16) and (6.9.17), that

$$\lambda(\boldsymbol{\phi}) = J \sum_{i=1}^{d} \cos\phi_i. \tag{6.10.7}$$

It follows that the behaviour of $R(\zeta)$ as $\zeta \to \zeta_c+$ is governed by the behaviour of the integrand of eqn (6.10.5) in the neighbourhood of the origin $\boldsymbol{\phi} = \mathbf{0}$. Thus, if we expand eqn (6.10.7) about the origin and transform to d-dimensional polar coordinates so that

$$\sum_{i=1}^{d} \phi_i^2 = r^2, \quad d\phi_1 \cdots d\phi_d = r^{d-1}\, dr d\Omega \tag{6.10.8}$$

where Ω denotes the solid angle, we obtain

$$R(\zeta) \sim C \int_0 \left(\varepsilon + \frac{J}{2} r^2\right)^{-1} r^{d-1}\, dr \qquad \text{as } \varepsilon \equiv \zeta - \zeta_c \to 0+. \tag{6.10.9}$$

In particular, when $\varepsilon = 0$, we see that $R(\zeta)$ diverges in dimensions one and two and is finite when $d > 2$. In other words, there is no phase transition in one and two dimensions when the interactions are between nearest neighbours only.

If we now look at the derivative of $R(\zeta)$, we have from eqns (6.10.5) and (6.10.9) that

$$R'(\zeta) \sim C \int_0 \left(\varepsilon + \frac{J}{2} r^2\right)^{-2} r^{d-1}\, dr \qquad \text{as } \varepsilon \to 0+ \tag{6.10.10}$$

and, in particular,

$$R'(\zeta_c) \sim C \int_0 r^{d-5}\, dr \qquad \text{is finite when } d > 4 \tag{6.10.11}$$

with a weak logarithmic divergence when $d = 4$. Treating d as a real variable in the range $2 < d < 4$, we obtain from eqn (6.10.10), after substituting $r = \varepsilon^{1/2} x$,

$$R'(\zeta) \sim D\varepsilon^{\frac{d}{2}-2} \quad \text{as } \varepsilon \to 0+ \qquad \text{for } 2 < d < 4. \tag{6.10.12}$$

Integrating eqn (6.10.12) and ignoring trivial multiplicative constants, we then have, on combining our previous results, the asymptotic forms

$$R(\zeta) - R(\zeta_c) \sim \begin{cases} \varepsilon^{(d-2)/2} & 2 < d < 4 \\ \varepsilon |\ln \varepsilon| & d = 4 \\ \varepsilon & d > 4 \end{cases} \qquad \text{as } \varepsilon \to 0+ \tag{6.10.13}$$

or, equivalently, since $R(\zeta)-R(\zeta_c)=\beta-\beta_c$,

$$\zeta-\zeta_c \sim \begin{cases} \tau^{2/(d-2)} & 2<d<4 \\ \tau|\ln \tau|^{-1} & d=4 \\ \tau & d>4 \end{cases} \quad \text{as } \tau=\frac{T}{T_c}-1\to 0+. \tag{6.10.14}$$

It is now an easy matter to deduce from eqns (6.10.4) and (6.10.14) that the zero-field susceptibility has the form

$$\chi_0 = \begin{cases} \tau^{-2/(d-2)} & 2<d<4 \\ \tau^{-1}\ln\tau & d=4 \\ \tau^{-1} & d\geqslant 4 \end{cases} \quad \text{as } \tau\to 0+ \tag{6.10.15}$$

from which one obtains the critical exponent

$$\gamma = \begin{cases} 2/(d-2) & 2<d<4 \\ 1 & d\geqslant 4. \end{cases} \tag{6.10.16}$$

Also, from eqns (6.10.6) and (6.10.14), the specific heat has the form

$$\frac{2}{k}C_0 \sim \begin{cases} 1-A\tau^{(4-d)/(d-2)} & 2<d<4 \\ 1-A(\ln\tau)^{-2} & d=4 \\ 1-A & d>4 \end{cases} \quad \text{as } \tau\to 0+ \tag{6.10.17}$$

where A is some positive constant depending on d. It follows that C_0 is continuous with a cusp singularity at T_c for $2<d\leqslant 4$ and has a jump discontinuity for $d>4$, so that from the definition of the critical exponent α for C_0, $\alpha=0$ for all $d>2$. To be consistent with scaling, however, we should interpret the exponent in eqn (6.10.17) for $2<d<4$ as the exponent for the singular part of the free energy $(2-\alpha)$ minus two. With this interpretation we then have

$$\alpha = \begin{cases} (d-4)/(d-2) & 2<d\leqslant 4 \\ 0_{\text{disc}} & d>4 \end{cases}. \tag{6.10.18}$$

Equations (6.10.16) and (6.10.18) and the universal result $\beta=1/2$ are then in accord with the scaling relation $\alpha+2\beta+\gamma=2$ for all $d>2$.

Finally, to obtain the exponent δ which characterizes the critical isotherm, we return to the saddle-point equation (6.9.27) and use the expression given in eqn (6.10.1) for the magnetization to eliminate ζ and obtain the equation of state

$$\beta = R(\zeta_c+H/2m)+\beta m^2 \tag{6.10.19}$$

or, equivalently,

$$\beta^{-1}[R(\zeta_c)-R(\zeta_c+H/2m)] = \tau+m^2 \qquad \left(\tau=\frac{T}{T_c}-1\right). \tag{6.10.20}$$

The asymptotic analysis of the left-hand side of eqn (6.10.20) is precisely the same as that given above in eqns (6.10.10)–(6.10.13) yielding

$$H \sim \begin{cases} m(\tau+m^2)^{2/(d-2)} & 2<d<4 \\ -m(\tau+m^2)[\ln(\tau+m^2)]^{-1} & d=4 \\ m(\tau+m^2) & d>4. \end{cases} \quad \text{as } h \to 0+ \quad (6.10.21)$$

In particular, on the critical isotherm $\tau = 0$, we obtain the exponent values

$$\delta = \begin{cases} (d+2)/(d-2) & 2<d<4 \\ 3 & d \geqslant 4. \end{cases} \quad (6.10.22)$$

Again, the above exponents satisfy the second scaling relation $\alpha + \beta(1+\delta) = 2$.

Notice also from the above, that the critical exponents for the spherical model take on the classical values $\alpha = 0_{\text{disc}}$, $\gamma = 1$, $\delta = 3$ for $d \geqslant 4$, while $\beta = 1/2$ has the classical value for any spherical model which has a phase transition. The cross-over to classical critical exponents at dimension four could well be a general feature of critical phenomena. Some critical exponents, for example γ for the Ising model, are in fact known rigorously (Aizenman 1981) to become classical for $d > 4$ and renormalization group arguments also predict this cross-over phenomenon for systems with short-range interactions.

For more general long-range power-law potentials of the form

$$J(\mathbf{r}) = J/|\mathbf{r}|^{d+\sigma} \qquad (J>0), \quad (6.10.23)$$

the Fourier transform (6.9.9) has, roughly speaking, the asymptotic form

$$\lambda(\boldsymbol{\phi}) - \lambda(\mathbf{0}) \sim \int^{\infty} (J/r^{d+\sigma})\, \mathrm{e}^{\mathrm{i}\phi r}\, r^{d-1}\, \mathrm{d}r$$

$$\sim A\phi^{\sigma} \qquad \text{as } \phi = |\boldsymbol{\phi}| \to 0+ \quad (6.10.24)$$

where in the second step, we have made the substitution $r = x\phi^{-1}$ and the constant A, in precise terms, is independent of ϕ and finite, only so long as $0 < \sigma < \min\{2, d\}$. Outside this range, σ is replaced by two and one recovers the short-range results above (see Joyce 1972).

If one works through the above analysis, assuming the asymptotic form eqn (6.10.24), the exponents for the power-law potential (6.10.23), can be obtained from the finite-range results by replacing $d/2$ by d/σ. With this replacement one obtains the critical exponents

$$\alpha = \begin{cases} (d-2\sigma)/(d-\sigma) & 1 < d/\sigma < 2 \\ 0_{\text{disc}} & d/\sigma \geqslant 2 \end{cases} \quad (6.10.25)$$

$$\gamma = \begin{cases} \sigma/(d-\sigma) & 1 < d/\sigma < 2 \\ 1 & d/\sigma \geqslant 2 \end{cases} \quad (6.10.26)$$

$$\delta = \begin{cases} (d+\sigma)/(d-\sigma) & 1 < d/\sigma < 2 \\ 3 & d/\sigma \geqslant 2 \end{cases} \qquad (6.10.27)$$

provided $0 < \sigma < \min\{2, d\}$. When $\sigma > \min\{2, d\}$, the exponents take on their short-range values and in all cases $\beta = 1/2$ whenever there is a phase transition.

Notice that for the power-law potential (6.10.23) the cross-over to classical behaviour occurs when $d/\sigma = 2$ and $\sigma < \min\{2, d\}$ and that in the non-classical domain the critical exponents depend continuously on a parameter appearing in the interaction potential.

With some degree of effort, correlation functions for the spherical model can be calculated and their asymptotic forms rigorously determined for large separations and in the neighbourhood of the critical point. The interested reader is referred to Joyce (1972) for details.

6.11 Problems

1. Use the transfer-matrix method to compute the canonical partition function for the following lattice models in one dimension with nearest-neighbour interactions only.
(a) *The spin-one Ising model.* Lattice sites are occupied by spins S_i which can take values 1, 0, or -1. In a given configuration the interaction energy is

$$E\{S\} = -\sum_{1 \leqslant i < j \leqslant N} J_{ij} S_i S_j.$$

(b) *The q-state Potts model* (Potts 1952). Lattice sites are assigned variables σ_i which take q distinct values. In a given configuration the interaction energy is

$$E\{\sigma\} = -\sum_{1 \leqslant i < j \leqslant N} J_{ij} \delta_{\sigma_i, \sigma_j}$$

where δ is the Kronecker symbol

$$\delta_{\sigma, \sigma'} = \begin{cases} 1 & \text{if } \sigma = \sigma' \\ 0 & \text{if } \sigma \neq \sigma'. \end{cases}$$

(c) *The Ashkin–Teller model* (Ashkin and Teller 1943). Lattice sites are assigned two Ising-spin variables $\mu_i = \pm 1$ and $\tau_i = \pm 1$. In a given configuration the interaction energy is

$$E\{\mu, \tau\} = -\sum_{1 \leqslant i < j \leqslant N} (J_{ij} \mu_i \mu_j + J'_{ij} \tau_i \tau_j + I_{ij} \mu_i \tau_i \mu_j \tau_j).$$

2. Compute the grand-canonical partition function $Z_G(V, T, z)$ for the one-dimensional lattice gas with interaction energy

$$E\{t\} = -J \sum_{i=1}^{V} t_i t_{i+1}$$

where $t_i = 0$ or 1, $t_{V+1} = t_1$, and

$$\sum_{i=1}^{V} t_i = N.$$

Show that for $J > 0$ the zeros of $Z_G(V, T, z)$ lie on the circle $|z| = e^{-K}$ in the complex z-plane, where $K = \beta J$.

3. Compute the free energy for the one-dimensional Ising model with nearest- and next-nearest-neighbour interactions having interaction energy

$$E\{\mu\} = -J_1 \sum_{i=1}^{N} \mu_i \mu_{i+1} - J_2 \sum_{i=1}^{N} \mu_i \mu_{i+2}$$

where $\mu_{N+i} \equiv \mu_i$ for $i = 1, 2$.

4. The canonical partition function for the one-dimensional n-vector model when $n = 2$ (the plane rotor or planar classical Heisenberg model) and with nearest-neighbour interactions only can be written as

$$Z_N = \int_0^{2\pi} \cdots \int \exp\left[K \sum_{i=1}^{N} \cos(\theta_{i+1} - \theta_i) \right] d\theta_1 \cdots d\theta_N$$

where $\theta_{N+1} \equiv \theta_1$.

Use a suitable modification of the transfer-matrix method to show that

$$Z_N = \sum_{i=1}^{\infty} \lambda_i^N$$

where $\lambda_1, \lambda_2, \ldots$ are the eigenvalues of the integral equation

$$\int_0^{2\pi} \exp[K \cos(x - y)]\, \phi(y) dy = \lambda \phi(x).$$

Compute the largest eigenvalue and the free energy per spin in the thermodynamic limit.

5. Consider a $2 \times N$ Ising lattice with row configurations $\mu = (\mu_1, \mu_2, \ldots, \mu_N)$, $\mu' = (\mu'_1, \mu'_2, \ldots, \mu'_N)$ and interaction energy

$$E\{\mu, \mu'\} = -J \sum_{i=1}^{N} \mu_i \mu'_i - J \sum_{i=1}^{N} (\mu_i \mu_{i+1} + \mu'_i \mu'_{i+1}).$$

Use the transfer-matrix method to show that the canonical partition function can be written as

$$Z_{2,N} = \sum_{\{\mu, \mu'\}} \exp(-\beta E\{\mu, \mu'\}) = \mathrm{tr}(\mathbb{A}^N)$$

where

$$\mathbb{A} = \begin{pmatrix} e^{3K} & 1 & 1 & e^{-K} \\ 1 & e^{K} & e^{-3K} & 1 \\ 1 & e^{-3K} & e^{K} & 1 \\ e^{-K} & 1 & 1 & e^{3K} \end{pmatrix}$$

and $K = \beta J$.

6. By performing an appropriate change of variables in eqn (6.4.11) for the internal energy E of the two-dimensional Ising model, derive eqn (6.4.17) for E in terms of the complete elliptic integral $K(k)$ of the first kind defined by eqn (6.4.19).

7. The complete elliptic integral $E(k)$ of the second kind is defined by eqn (6.4.21). From the definitions of $E(k)$ and $K(k)$, show that

$$k\frac{\mathrm{d}}{\mathrm{d}k}K(k) = (1-k^2)^{-1}E(k) - K(k).$$

Use this result to derive eqn (6.4.20) for the specific heat C_0 of the two-dimensional Ising model.

8. By writing

$$K(k) = \int_0^{\pi/2} (1-k\sin\theta)(1-k^2\sin^2\theta)^{-1/2}\,\mathrm{d}\theta + \int_0^{\pi/2} k\sin\theta(1-k^2\sin^2\theta)^{-1/2}\,\mathrm{d}\theta$$

show that

$$K(k) \sim \ln[4(1-k^2)^{-1/2}] \qquad \text{as } k\to 1-.$$

9. Show that the equation of state for the one-dimensional hard-core lattice gas is

$$\beta p = \ln\tfrac{1}{2}[1+(1-2\rho)^{-1}].$$

10. Show that the free energy of the one-dimensional spherical model with nearest-neighbour interactions only is given by

$$-\beta\psi(\beta,0) = \frac{1}{2}\{(1+4K^2)^{1/2} - \ln\frac{1}{4\pi}[1+(1+4K^2)^{1/2}]\}$$

where $K = \beta J$ and J is the nearest-neighbour coupling constant.

11. Show, using eqn (3.5.13), that the result of Problem 10 can be obtained by taking an appropriate $n\to\infty$ limit of a sequence of n-vector models.

12. Derive the general expression given in eqn (6.10.6) for the zero-field specific heat of the spherical model.

6.12 Solutions to problems

1. In general, the interaction energy of a one-dimensional lattice model with nearest-neighbour interactions only can be expressed in the form

$$E\{\sigma\} = \sum_{i=1}^{N} V(\sigma_i, \sigma_{i+1})$$

where σ_i represents the 'state' of site i and, for simplicity, we have assumed periodic boundary conditions ($\sigma_{N+1} = \sigma_1$).

For such models the canonical partition function Z_N can be written in the form

$$Z_N = \sum_{i=1}^{M} \lambda_i^N$$

where, from eqns (6.2.3)–(6.2.6), λ_i, $i = 1, 2, \ldots, M$ are the eigenvalues of the transfer matrix $\mathbb{T}$ with elements

$$T(\sigma,\sigma') = \exp[-\beta V(\sigma,\sigma')].$$

(a) For the spin-one Ising chain the site state variable is $S = 1, 0, -1$ and

$$V(S, S') = -JSS'.$$

The transfer matrix

$$\begin{array}{c} S\backslash S' \\ \\ \mathbb{T} = \\ \\ \end{array} \begin{array}{c} \\ 1 \\ 0 \\ -1 \end{array} \begin{array}{c} \begin{array}{ccc} 1 & 0 & -1 \end{array} \\ \begin{pmatrix} e^{K} & 1 & e^{-K} \\ 1 & 1 & 1 \\ e^{-K} & 1 & e^{K} \end{pmatrix} \end{array}$$

has eigenvalues

$$\lambda_1, \lambda_2 = \tfrac{1}{2}\{1 + 2\cosh K \pm [(2\cosh K - 1)^2 + 8]^{1/2}\}$$

$$\lambda_3 = 2 \sinh K$$

where $K = \beta J$ and, in the thermodynamic limit, the free energy per site is given by

$$-\beta\psi = \ln\tfrac{1}{2}\{1 + 2\cosh K + [(2\cosh K - 1)^2 + 8]^{1/2}\}.$$

(b) For the q-state Potts model,

$$V(\sigma, \sigma') = -J\delta_{\sigma,\sigma'}; \qquad \sigma, \sigma' = 1, 2, 3, \ldots, q.$$

The transfer matrix

$$\mathbb{T} = \begin{array}{c} \sigma\backslash\sigma' \\ 1 \\ 2 \\ 3 \\ \vdots \\ q \end{array} \begin{array}{c} \begin{array}{ccccc} 1 & 2 & 3 & \cdots & q \end{array} \\ \begin{pmatrix} e^{K} & 1 & 1 & \cdots & 1 \\ 1 & e^{K} & 1 & \cdots & 1 \\ 1 & 1 & e^{K} & \cdots & 1 \\ & & & & \\ 1 & 1 & 1 & \cdots & e^{K} \end{pmatrix} \end{array}$$

is a symmetric cyclic matrix with elements of the form $T_{ij} = T(i-j) = T_{ji}$. In general, such matrices are diagonalized by the matrix $\mathbb{S}$ with elements

$$S_{kl} = q^{-1/2} \exp(2\pi i kl/q).$$

That is,

$$(\mathbb{S}^{-1}\,\mathbb{T}\mathbb{S})_{kl} = \delta_{k,l}\,\lambda\!\left(\frac{2\pi k}{q}\right) \qquad k = 1, 2, \ldots, q$$

where

$$\lambda(\phi) = \sum_{s=0}^{q-1} T(s) \exp(is\phi).$$

In the present instance the maximum eigenvalue is

$$\lambda(2\pi) = e^{K} + q - 1.$$

The remaining eigenvalues $\lambda(2\pi k/q)$, $k = 1, 2, \ldots, q-1$ are $(q-1)$-fold degenerate and take the value $e^{K} - 1$.

The free energy per site in the thermodynamic limit is given by

$$-\beta\psi = \ln(e^{K} + q - 1).$$

(c) For the Ashkin–Teller model the state variable for site i is the pair $\sigma_i = (\mu_i, \tau_i)$ and

$$V(\sigma, \sigma') = -J\mu\mu' - J'\tau\tau' - I\mu\tau\mu'\tau'.$$

The transfer matrix ($K = \beta J$, $K' = \beta J'$, $L = \beta I$)

$$\mathbb{T} = \begin{array}{c|cccc} \sigma\backslash\sigma' & (1,1) & (1,-1) & (-1,1) & (-1,-1) \\ \hline (1,1) & e^{K+K'+L} & e^{K-K'-L} & e^{-K+K'-L} & e^{-K-K'+L} \\ (1,-1) & e^{K-K'-L} & e^{K+K'+L} & e^{-K-K'+L} & e^{-K+K'-L} \\ (-1,1) & e^{-K+K'-L} & e^{-K-K'+L} & e^{K+K'+L} & e^{K-K'-L} \\ (-1,-1) & e^{-K-K'+L} & e^{-K+K'-L} & e^{K-K'-L} & e^{K+K'+L} \end{array}$$

has a very simple structure with eigenvalues

$$e^{K+K'+L} + e^{K-K'-L} + e^{-K+K'-L} + e^{-K-K'+L}$$
$$e^{K+K'+L} - e^{K-K'-L} + e^{-K+K'-L} - e^{-K-K'+L}$$
$$e^{K+K'+L} - e^{K-K'-L} - e^{-K+K'-L} + e^{-K-K'+L}$$
$$e^{K+K'+L} + e^{K-K'-L} - e^{-K+K'-L} - e^{-K-K'+L}$$

and corresponding eigenvectors

$$\begin{pmatrix} 1 \\ 1 \\ 1 \\ 1 \end{pmatrix}, \quad \begin{pmatrix} 1 \\ -1 \\ 1 \\ -1 \end{pmatrix}, \quad \begin{pmatrix} 1 \\ -1 \\ -1 \\ 1 \end{pmatrix}, \quad \begin{pmatrix} 1 \\ 1 \\ -1 \\ -1 \end{pmatrix}.$$

The free energy per site in the thermodynamic limit is given by

$$-\beta\psi = \ln 2[e^K \cosh(K' + L) + e^{-K} \cos(K' - L)].$$

2. Using the transfer matrix formalism, the grand-canonical partition function can be expressed in the form

$$Z_G(V, T, z) = \sum_{\{t\}} \exp\left(\ln z \sum_{k=1}^{V} t_k\right) \exp\left(K \sum_{k=1}^{V} t_k t_{k+1}\right)$$
$$= \mathrm{tr}(\mathbb{T}^V)$$

where $K = \beta J$ and the transfer matrix $\mathbb{T}$ has elements

$$T(t, t') = z^{t/2} e^{Ktt'} z^{t'/2}; \qquad t, t' = 0, 1.$$

That is,

$$\mathbb{T} = \begin{pmatrix} 1 & z^{1/2} \\ z^{1/2} & ze^K \end{pmatrix}.$$

The eigenvalues of $\mathbb{T}$ are

$$\lambda_1, \lambda_2 = \tfrac{1}{2}\{1 + ze^K \pm [(ze^K - 1)^2 + 4z]^{1/2}\} \equiv a \pm b$$

and

$$Z_G(V, T, z) = \lambda_1^V + \lambda_2^V.$$

Obviously, Z_G vanishes when

$$\lambda_2/\lambda_1 = \exp[(2k+1)i\pi/V], \qquad k = 0, 1, 2, \ldots, V-1. \tag{1}$$

Now, since $a^2 - b^2 = z(e^K - 1)$, we can parametrize the problem by writing

$$a = z^{1/2}(e^K - 1)^{1/2} \cosh \psi, \qquad b = z^{1/2}(e^K - 1)^{1/2} \sinh \psi.$$

We then have

$$\lambda_2/\lambda_1 = (a-b)(a+b)^{-1} = \exp(-2\psi).$$

Comparison with eqn (1) then shows that Z_G vanishes when

$$a = \tfrac{1}{2}(1 + ze^K) = z^{1/2}(e^K - 1)^{1/2} \cos[(2k+1)\pi/2V]$$

$$k = 0, 1, \ldots, V-1.$$

Solving for z, we find that the zeros of Z_G are given by

$$z_j = e^{-K}e^{\pm i\theta_j}, \qquad j = 0, 1, \ldots, V-1$$

where

$$\cos\theta_j = -e^{-K} + (1 - e^{-K})\cos[(2j-1)\pi/V]$$

which gives the required result.

Note this is an explicit example of the celebrated Lee–Yang circle theorem (Lee and Yang 1952).

3. The canonical partition function is given by

$$Z_N = \sum_{\{\mu\}} \exp\left(K_1 \sum_{i=1}^{N} \mu_i\mu_{i+1} + K_2 \sum_{i=1}^{N} \mu_i\mu_{i+2}\right).$$

If we define new Ising spin variables τ_i by $\tau_i = \mu_i\mu_{i+1}$, the sum over $\{\mu\}$ can be replaced by a sum over $\{\tau\}$ and, since $(\mu_{i+1})^2 = 1$, we have

$$Z_N = \sum_{\{\tau\}} \exp\left[K_1 \sum_{i=1}^{N} \tau_i + K_2 \sum_{i=1}^{N} \tau_i\tau_{i+1}\right]$$

which is equivalent to the nearest-neighbour Ising chain problem in an external field.

The free energy, from eqn (6.3.7), is given by

$$-\beta\psi = \ln[e^{K_2}\cosh K_1 + (e^{2K_2}\sinh^2 K_1 + e^{-2K_2})^{1/2}].$$

4. If we define the kernel

$$K(\theta, \psi) = \exp[K\cos(\theta - \psi)],$$

the partition function for the plane rotor chain can be written as

$$Z_N = \int_0^{2\pi}\!\!\cdots\!\int K(\theta_1, \theta_2)K(\theta_2, \theta_3)\cdots K(\theta_{N-1}, \theta_N)\, K(\theta_N, \theta_1)d\theta_1 \cdots d\theta_N. \tag{1}$$

It is not difficult to show (since K is a so-called Hilbert–Schmidt kernel) that the integral equation

$$\int_0^{2\pi} K(x, y)\psi(y)\,dy = \lambda\psi(x)$$

has a discrete set of eigenvalues λ_i and corresponding eigenfunctions ψ_i which form a complete orthonormal set. The kernel K can then be expressed as a convergent series

$$K(x, y) = \sum_{i=1}^{\infty} \lambda_i \psi_i(x) \psi_i(y) \tag{2}$$

(otherwise known as the spectral theorem).

Substituting eqn (2) into eqn (1) and using the orthonormality of the eigenfunctions,

$$\int_0^{2\pi} \psi_i(x)\psi_j(x)\mathrm{d}x = \delta_{i,j},$$

gives the required result.

5. If we define $\sigma_i = (\mu_i, \mu_i')$, the canonical partition function can be expressed in the form

$$\begin{aligned} Z_{2,N} &= \sum_{\{\mu\}} \exp\left(K \sum_{i=1}^{N} \mu_i\mu_i' + \mu_i\mu_{i+1} + \mu_i'\mu_{i+1}' \right) \\ &= \sum_{\{\sigma\}} \prod_{i=1}^{N} A(\sigma_i, \sigma_{i+1}) \\ &= \mathrm{tr}(\mathbb{A}^N) \end{aligned}$$

where, for $\sigma = (\mu, \mu')$ and $\sigma' = (\tau, \tau')$,

$$A(\sigma, \sigma') = \exp(K\mu\mu'/2) \exp[K(\mu\tau + \mu'\tau')] \exp(K\tau\tau'/2)$$

or, in matrix form,

$$\begin{array}{cc} & \begin{array}{cccc} (1,1) & (-1,1) & (1,-1) & (-1,-1) \end{array} \\ \mathbb{A} = \begin{array}{c} \sigma\backslash\sigma' \\ (1,1) \\ (-1,1) \\ (1,-1) \\ (-1,-1) \end{array} & \left[\begin{array}{cccc} \mathrm{e}^{3K} & 1 & 1 & \mathrm{e}^{-K} \\ 1 & \mathrm{e}^{K} & \mathrm{e}^{-3K} & 1 \\ 1 & \mathrm{e}^{-3K} & \mathrm{e}^{K} & 1 \\ \mathrm{e}^{-K} & 1 & 1 & \mathrm{e}^{3K} \end{array}\right]. \end{array}$$

6. Changing variables to

$$\omega_1 = \tfrac{1}{2}(\theta_1 + \theta_2), \qquad \omega_2 = \tfrac{1}{2}(\theta_1 - \theta_2)$$

in the integral appearing in eqn (6.4.11), we have

$$E = -J \coth 2K[1 + (2\tanh^2 2K - 1) I(k_1)]$$

where

$$\begin{aligned} I(k) &= \pi^{-2} \iint_0^{\pi} \left[1 - \frac{k}{2}(\cos\theta_1 + \cos\theta_2) \right]^{-1} \mathrm{d}\theta_1\, \mathrm{d}\theta_2 \\ &= \pi^{-2} \iint_0^{\pi} [1 - k \cos\omega_1 \cos\omega_2]^{-1} \mathrm{d}\omega_1 \mathrm{d}\omega_2 \end{aligned}$$

and

$$k_1 = 2 \sinh 2K / \cosh 2K.$$

The required result then follows almost immediately by noting that

$$\pi^{-1}\int_0^{\pi}(1-a\cos\omega)^{-1}\,\mathrm{d}\omega=(1-a^2)^{-1/2}$$

and

$$K(k)=\int_0^{\pi/2}(1-k^2\sin^2\theta)^{-1/2}\,\mathrm{d}\theta=\frac{1}{2}\int_0^{\pi}(1-k^2\cos^2\theta)^{-1/2}\,\mathrm{d}\theta.$$

7. By definition,

$$k\frac{\mathrm{d}K(k)}{\mathrm{d}k}=\int_0^{\pi/2}k^2\sin^2\theta(1-k^2\sin^2\theta)^{-3/2}\,\mathrm{d}\theta=\int_0^{\pi/2}(1-k^2\sin^2\theta)^{-3/2}-K(k). \quad (1)$$

Expanding the integrand in eqn (1) in a uniformly convergent ($k^2<1$) series and integrating term-by-term, we have

$$(1-k^2)\int_0^{\pi/2}(1-k^2\sin^2\theta)^{-3/2}\,\mathrm{d}\theta$$

$$=\frac{\pi}{2}+\sum_{n=1}^{\infty}\frac{\Gamma\left(-\frac{1}{2}\right)(-1)^n k^{2n}}{n!\,\Gamma\left(-\frac{1}{2}-n\right)}\left(I_n-\frac{2n}{2n+1}I_{n-1}\right) \quad (2)$$

where $\Gamma(z)$ is the gamma function and

$$I_n=\int_0^{\pi/2}\sin^{2n}\theta\,\mathrm{d}\theta=2\pi(2n)!/4^{n+1}(n!)^2=\left(1-\frac{1}{2n}\right)I_{n-1}. \quad (3)$$

Using eqn (3) and the fact that $\Gamma(z+1)=z\Gamma(z)$, we obtain

$$\int_0^{\pi/2}(1-k^2\sin^2\theta)^{-3/2}\,\mathrm{d}\theta=(1-k^2)^{-1}$$

$$\times\left[\frac{\pi}{2}+\sum_{n=1}^{\infty}\frac{4\Gamma\left(\frac{3}{2}\right)(-1)^n k^{2n}I_n}{n!\,\Gamma\left(-\frac{1}{2}-n\right)(2n+1)(2n-1)}\right]$$

$$=(1-k^2)^{-1}\int_0^{\pi/2}\sum_{n=0}^{\infty}\frac{\Gamma\left(\frac{3}{2}\right)(-1)^n k^{2n}}{n!\,\Gamma\left(\frac{3}{2}-n\right)}\sin^{2n}\theta\,\mathrm{d}\theta$$

$$=(1-k^2)^{-1}\int_0^{\pi/2}(1-k^2\sin^2\theta)^{1/2}\,\mathrm{d}\theta$$

$$=(1-k^2)^{-1}E(k). \quad (4)$$

Substituting eqn (4) into eqn (1) gives the required result.

The eqn (6.4.20) for C_0 follows easily by differentiating the expression given in eqn (6.4.17) for E with respect to T and using the above result.

8. Changing variables to $x = \cos\theta$,

$$\begin{aligned} I_1 &= \int_0^{\pi/2} k\sin\theta(1-k^2\sin^2\theta)^{-1/2}\,d\theta \\ &= \int_0^1 k(1-k^2+k^2x^2)^{-1/2}\,dx \\ &= [\ln\{kx+(1-k^2+k^2x^2)^{1/2}\}]_0^1 \\ &= \ln[(1+k)(1-k^2)^{-1/2}]. \end{aligned}$$

Also, as $k \to 1-$,

$$\begin{aligned} I_2 &= \int_0^{\pi/2} (1-k\sin\theta)(1-k^2\sin^2\theta)^{-1/2}\,d\theta \\ &= \int_0^{\pi/2} [(1-k\cos\theta)/(1+k\cos\theta)]^{1/2}\,d\theta \\ &\sim \int_0^{\pi/2} \tan(\theta/2)\,d\theta = -[2\ln\cos(\theta/2)]_0^{\pi/2} \\ &= \ln 2. \end{aligned}$$

It follows from the above results that

$$K(k) = I_1 + I_2 \sim \ln[4(1-k^2)^{-1/2}] \qquad \text{as } k \to 1-.$$

9. The grand-canonical partition function for the one-dimensional hard-core lattice gas can be obtained from Problem 2 by taking the limit $K \to \infty$.
In the thermodynamic limit

$$\begin{aligned} \chi(z) &= \lim_{V\to\infty} V^{-1}\ln Z_G(V,z) \\ &= \ln\lambda_1 \\ &= \ln\tfrac{1}{2}[1+(1+4z)^{1/2}]. \end{aligned}$$

The density is given by

$$\rho = z\frac{\partial\chi}{\partial z} = \frac{1}{2}[1-(1+4z)^{-1/2}].$$

Solving for z in terms of ρ gives the equation of state

$$\begin{aligned} \beta p &= \ln\tfrac{1}{2}\{1+[1+4z(\rho)]^{1/2}\} \\ &= \ln\tfrac{1}{2}[1+(1-2\rho)^{-1}]. \end{aligned}$$

10. The free energy of the one-dimensional spherical model with nearest-neighbour coupling J only is given, from eqns (6.9.17) and (6.9.26), by

$$\begin{aligned} -\beta\psi(\beta,0) &= Kz - \frac{1}{2}\ln(K/2) - (4\pi)^{-1}\int_0^{2\pi} \ln 2(z-\cos\theta)\,d\theta + \frac{1}{2}\ln\pi \\ &= Kz - \tfrac{1}{2}\ln(K/2\pi) - \tfrac{1}{2}\ln[z+(z^2-1)^{1/2}] \qquad (1) \end{aligned}$$

where $K = \beta J$ and $z = z(K)$ is determined from the saddle-point condition (6.9.27),

$$K = (4\pi)^{-1}\int_0^{2\pi}(z-\cos\theta)^{-1}\,d\theta = \frac{1}{2}(z^2-1)^{-1/2} \tag{2}$$

Solving eqn (2) for z and substituting into eqn (1) gives

$$-\beta\psi(\beta, 0) = \frac{1}{2}\left\{(1+4K^2)^{1/2} - \ln\frac{1}{4\pi}[1+(1+4K^2)^{1/2}]\right\}$$

11. In order to proceed to the limit $n \to \infty$, it is customary to normalize the partition function by dividing by the volume of the phase space which is given from Problem 1 of Chapter 2 by

$$(\int d\sigma)^N = [2\pi^{n/2}/\Gamma(n/2)]^N.$$

The normalized partition function, with the nearest-neighbour coupling constant scaled by a factor n, is then given by

$$Z_n(nK) = [\lambda(nK)]^{N-1} \tag{1}$$

where, from eqns (3.5.12) and (3.5.13),

$$\lambda(nK) = \Gamma(n/2)\left(\frac{nK}{2}\right)^{1-\frac{n}{2}} I_{\frac{n}{2}-1}(nK)$$

$$= \Gamma(n/2)\,(2\pi i)^{-1}\int_{c-i\infty}^{c+i\infty}\exp[p+(nK)^2/4p]p^{-\frac{n}{2}}\,dp.$$

Changing variables to $p = (nK/2)z$, we then obtain

$$\lambda(nK) = \Gamma(n/2)\,(nK/2)^{1-\frac{n}{2}}\,(2\pi i)^{-1}\int_{\varepsilon-i\infty}^{\varepsilon+i\infty}\exp\left[\frac{n}{2}f(z)\right]dz \tag{2}$$

where

$$f(z) = K(z+z^{-1}) - \ln z$$

and ε is any positive number.

For large n the integral in eqn (2) can be evaluated by the method of steepest descents.

The stationary points of $f(z)$ occur when

$$f'(z) = K(1-z^{-2}) - z^{-1} = 0.$$

That is

$$z = (2K)^{-1}[1 \pm (1+4K^2)^{1/2}].$$

Since $\varepsilon > 0$ in eqn (2), we deform the contour through the saddle point

$$z_0 = (2K)^{-1}[1+(1+4K^2)^{1/2}].$$

Using Stirling's formula $\Gamma(n/2) \sim (n/2)^{n/2}\exp(-n/2)$, we then arrive at the asymptotic formula

$$\lambda(nK) \sim \exp\frac{n}{2}[f(z_0) - 1 - \ln K] \qquad \text{as } n \to \infty$$

and finally, after taking the thermodynamic limit, we obtain from eqn (1) the limiting free energy

$$-\beta\psi = \lim_{N,n\to\infty} (Nn)^{-1} \ln Z_N(nK)$$
$$= \tfrac{1}{2}\{(1+4K^2)^{1/2} - 1 - \ln\tfrac{1}{2}[1+(1+4K^2)^{1/2}]\}$$

which, apart from trivial normalization factors, is identical with the spherical model result given in eqn (6.6.10).

In fact, if one normalizes the spherical model partition function by the surface area $2\pi^{N/2}N^{(N-1)/2}/\Gamma(N/2)$ of the N-dimensional sphere of radius $N^{1/2}$ the results of Problems 10 and 11 are in precise agreement.

12. The zero-field free energy for the spherical model is given in general, from eqn (6.9.26), by

$$-\beta\psi(\beta, 0) = \beta\zeta - \frac{1}{2}(2\pi)^{-d}\int_0^{2\pi}\cdots\int \ln \pi\beta[\zeta - \lambda(\boldsymbol{\phi})]\,d\phi_1 \cdots d\phi_d \tag{1}$$

where $\zeta = \zeta(\beta)$ is determined from the saddle-point condition (6.9.27)

$$\beta = \frac{1}{2}(2\pi)^{-d}\int_0^{2\pi}\cdots\int [\zeta - \lambda(\boldsymbol{\phi})]^{-1}\,d\phi_1 \cdots d\phi_d. \tag{2}$$

The zero-field specific heat is given, in general, by

$$C_0 = -T\frac{\partial^2\psi}{\partial T^2}(\beta, 0) = k\beta^2\frac{\partial^2}{\partial\beta^2}(-\beta\psi).$$

From eqns (1) and (2),

$$\frac{\partial}{\partial\beta}(-\beta\psi) = \zeta - (2\beta)^{-1} + \frac{\partial\zeta}{\partial\beta}\left\{\beta - \frac{1}{2}(2\pi)^{d}\int_0^{2\pi}\cdots\int [\zeta - \lambda(\boldsymbol{\phi})]^{-1}\,d\phi_1 \cdots d\phi_d\right\}$$
$$= \zeta - (2\beta)^{-1}.$$

It then follows that

$$C_0 = k/2 + k\beta^2\frac{\partial\zeta}{\partial\beta}$$

which gives the result stated in eqn (6.10.6) when one notes that for $T < T_c$ the saddle point ζ sticks at the (constant) value ζ_c.

7

SCALING THEORY AND THE RENORMALIZATION GROUP

7.1 Introduction

In Chapter 4, we introduced the idea of a critical exponent as a means of characterizing the nature of singular behaviour in the neighbourhood of a phase transition or critical point. Thus, for a function f with a critical point or singularity at $x = 0$, we define the critical exponent ω by

$$\omega = \lim_{x \to 0+} \ln|f(x)|/\ln|x| \tag{7.1.1}$$

whenever the limit exists, and we write

$$f(x) \sim x^{\omega} \qquad \text{as } x \to 0+. \tag{7.1.2}$$

Similarly, we define the exponent ω' as the left-sided limit $x \to 0-$ of eqn (7.1.1).

For magnetic systems we defined, in Table 4.1, the critical exponents $\alpha, \beta, \gamma, \delta$ corresponding to the critical point $\tau = (T/T_c) - 1 = 0$ by

$$\begin{aligned} &C_0 \sim \tau^{-\alpha}, \quad \chi_0 \sim \tau^{-\gamma} && \text{as } \tau \to 0+ \\ &m_0 \sim |\tau|^{\beta} && \text{as } \tau \to 0- \\ &m \sim H^{1/\delta} \quad \text{at } \tau = 0 && \text{as } H \to 0+ \end{aligned} \tag{7.1.3}$$

where C_0, χ_0, and m_0 are the zero-field specific heat, isothermal susceptibility, and spontaneous magnetization, respectively, m is the magnetization and H the magnetic field. The exponents α' and γ' defined previously are the low-temperature ($\tau \to 0-$) exponents for C_0 and χ_0.

The above exponents were also defined in Table 4.1 for fluid systems. We will confine ourselves in this chapter, however, to magnetic systems.

We pointed out in Section 4.5 that inequalities relating the above exponents appear to be satisfied as equalities. In particular, we have seen that relations

$$\alpha' + 2\beta + \gamma' = 2, \quad \alpha' + \beta(1 + \delta) = 2 \tag{7.1.4}$$

hold rigorously for the classical theories and the two-dimensional Ising model. Scaling theory predicts that these relations hold in general.

The basis for scaling arguments is the hypothesis that close to the critical point a change of scale for the deviation of temperature and magnetic field from their critical values by factors λ^{a_τ} and λ^{a_H} for real λ, is equivalent to a linear change of scale by a factor λ for the singular part of the free energy. Granted this hypothesis, we show in the following section that the critical exponents defined above can all be expressed in terms of a_τ and a_H. Elimination of a_τ and a_H results in the *scaling laws* (7.1.4).

The Kadanoff 'block-spin' picture of scaling is presented in Section 7.3 and is used to obtain scaling relations for the correlation function exponents defined in Section 5.4.

In renormalization-group approach to critical phenomena, due to Wilson (1971), builds on the block-spin idea of Kadanoff (1966) and provides a recursive or renormalization scheme for handling large fluctuations and correlation lengths in the vicinity of a critical point. The renormalization-group recipe is discussed in Section 7.4 and is shown to provide a firm basis for scaling theory and the so-called universality of critical behaviour. The renormalization scheme gives a method for actually calculating the scaling exponents a_τ and a_H defined above, from which one can then deduce the values of critical exponents which, depending on the renormalization scheme, fall naturally into certain universality classes.

While the renormalization-group recipe has considerable intuitive appeal, its actual implementation requires approximations which are often suspect and virtually impossible to control. Some exact and approximate examples of renormalization-group schemes are given in Section 7.5

7.2 The scaling hypothesis and critical-exponent relations

When a magnetic system has a phase transition, its free energy $\psi(T, H)$ is, by definition, singular in zero field at a certain critical temperature T_c. If we separate ψ into its regular and singular parts and consider the singular part ψ_s to be a function of $\tau = (T/T_c) - 1$ and H in the neighbourhood of its critical point, the scaling hypothesis in its simplest form, states that for any real λ there exist exponents a_τ and a_H such that

$$\psi_s(\lambda^{a_\tau}\tau, \lambda^{a_H}H) = \lambda\psi_s(\tau, H). \tag{7.2.1}$$

In other words, scaling the temperature τ and field H variables by factors λ^{a_τ} and λ^{a_H} in the neighbourhood of the critical point $\tau = H = 0$ is equivalent, for certain a_τ and a_H, to scaling the singular part of the free energy by a factor λ. Mathematically, eqn (7.2.1) states that ψ_s is a *generalized homogeneous function* of τ and H.

There are many versions of the scaling hypothesis, none of which has been justified, but all are in effect equivalent to the assumption stated in eqn (7.2.1). The block-spin version of Kadanoff (1966) will be discussed in the following section since it gives additional scaling properties for correlation functions and provides some background to the renormalization-group approach.

The earlier version of scaling due to Widom (1965) assumed that ψ_s could be written as a product of $|\tau|$ to some power and a function of a single scaled variable. To see how this follows from eqn (7.2.1) we use the fact that λ is arbitrary to make the particular choice $\lambda = |\tau|^{-1/a_\tau}$. For this value of λ, eqn (7.2.1) gives

$$\begin{aligned}\psi_s(\tau, H) &= |\tau|^{1/a_\tau}\, \psi_s(\pm 1, H/|\tau|^{\Delta}) \\ &\equiv |\tau|^{1/a_\tau}\, Q_{\pm}(H/|\tau|^{\Delta}) \qquad (7.2.2)\end{aligned}$$

where $\Delta = a_H/a_\tau$, and the '+' and '−' correspond, respectively, to high ($\tau > 0$) and low ($\tau < 0$) temperatures. Obviously, any function of the form given in eqn (7.2.2) is a generalized homogeneous function and vice versa.

If we now assume that the functions $Q_\pm$ and their derivatives are finite at the origin, we obtain, by definition and from eqn (7.2.2), the asymptotic forms

$$C_0 \sim -T\frac{\partial^2 \psi_s}{\partial T^2}(\tau, 0) \sim |\tau|^{1/a_\tau - 2} \qquad \text{as } \tau \to 0,$$

$$m_0 \sim -\left(\frac{\partial \psi_s}{\partial H}\right)_{H=0} \sim |\tau|^{1/a_\tau - \Delta} \qquad \text{as } \tau \to 0-, \qquad (7.2.3)$$

and

$$\chi_0 \sim -\left(\frac{\partial^2 \psi_s}{\partial H^2}\right)_{H=0} \sim |\tau|^{1/a_\tau - 2\Delta} \qquad \text{as } \tau \to 0.$$

Comparison with the definitions given in eqn (7.1.3) for the exponents α, β, and γ then shows that

$$\alpha = 2 - 1/a_\tau,$$

$$\beta = 1/a_\tau - \Delta, \qquad (7.2.4)$$

and

$$\gamma = 2\Delta - 1/a_\tau$$

from which the first scaling relation given in eqn (7.1.4) follows when it is noted that the above form of the scaling hypothesis implicitly assumes the exponent symmetry $\alpha' = \alpha$ and $\gamma' = \gamma$. Critical-exponent asymmetry can of course be incorporated in the above scheme by making different scaling hypotheses above and below T_c.

In order to obtain the second scaling relation given in eqn (7.1.4) we first note, from eqn (7.2.2), that the magnetization in non-zero field has the form

$$m \sim -\frac{\partial \psi_s}{\partial H} = -|\tau|^{1/a_\tau - \Delta}\, Q'_{\pm}(H/|\tau|^{\Delta}). \qquad (7.2.5)$$

In order to match eqn (7.2.5) to the definition of δ given in eqn (7.1.3) we require $Q'_\pm(x)$ to have the asymptotic form

$$Q'_{\pm}(x) \sim -x^{1/\delta} \qquad \text{as } x \to \infty. \qquad (7.2.6)$$

It then follows that, for $H > 0$ fixed,

$$m \sim |\tau|^{1/a_\tau - \Delta - \Delta/\delta} H^{1/\delta} \qquad \text{as } \tau \to 0 \tag{7.2.7}$$

and hence that

$$1/a_\tau - \Delta(1 + 1/\delta) = 0. \tag{7.2.8}$$

However, from eqn (7.2.4), we have

$$1/a_\tau = 2 - \alpha, \qquad \Delta = 2 - \alpha - \beta. \tag{7.2.9}$$

Substituting eqn (7.2.9) into eqn (7.2.8) gives, after some rearranging, the scaling relation

$$\alpha + \beta(1 + \delta) = 2. \tag{7.2.10}$$

As noted previously, the above scaling relations hold rigorously for the classical theories and the two-dimensional Ising model. Approximate numerical values for other model spin-system critical exponents, including, in particular, those of the three-dimensional Ising model, also appear to satisfy these relations although it is probably fair to say that the estimates are somewhat biased in this direction. Also, within the range of experimental error, values of critical exponents for real systems satisfy the above scaling relations. We stress, however, that there is no rigorous justification of the scaling hypothesis although it seems to be the case that the two relations derived above are pretty well satisfied both theoretically and experimentally.

7.3 Block-spins and scaling of the correlation functions

The idea of block-spins is due to Kadanoff (1966) and is most easily stated for Ising spin systems. Consider then a d-dimensional hypercubic lattice of Ising spins $\mu_{\mathbf{r}} = \pm 1$ located at lattice sites $\mathbf{r}$ and assume, for simplicity, that the lattice spacing is unity and that only nearest-neighbour spins interact. We divide the lattice up into cubic cells with sides of length L so that each cell contains L^d spins. We then define m_α to be the magnetization per site of the cell labelled α. That is

$$m_\alpha = L^{-d} \sum_{\mathbf{r} \in \alpha} \mu_{\mathbf{r}}. \tag{7.3.1}$$

Now, as we approach the critical point, the correlation length ξ becomes large and spins are correlated over large distances so that, if L is large compared with unity but small compared with ξ, the spins in each cell will tend to become aligned with one another and be either predominantly 'up' or 'down'. Kadanoff expresses this idea by defining the *block-spin variable* σ_α for cell α by

$$\sigma_\alpha = m_\alpha / \langle \mu \rangle_L \tag{7.3.2}$$

where $\langle\mu\rangle_L$, given by

$$\langle\mu\rangle_L^2 = L^{-2d} \sum_{\mathbf{r}\in\alpha,\mathbf{r}'\in\alpha} \langle\mu_{\mathbf{r}}\mu_{\mathbf{r}'}\rangle, \tag{7.3.3}$$

may be thought of as a measure of long-range order within cells.

From the above heuristic argument we assume that, close to the critical point, the new spin variables are also Ising spins $\sigma_\alpha = \pm 1$ and that, to a first approximation, only nearest-neighbour cells interact. In this way we map the original system, with temperature and field variables $\tau = (T/T_c) - 1$ and H, respectively, into itself with new variables $\tilde{\tau}$ and $\tilde{H}$. At the same time we reduce the number of degrees of freedom by a factor L^d and change the length scale by a factor L. Thus, if $\psi_s(\tau, H)$ denotes the singular part of the free energy, we have

$$\psi_s(\tau, H) = L^{-d}\psi_s(\tilde{\tau}, \tilde{H}). \tag{7.3.4}$$

Similarly, if we define the pair-correlation function by

$$G(\tau, H, r) = \langle\mu_{\mathbf{s}}\mu_{\mathbf{s}+\mathbf{r}}\rangle - \langle\mu_{\mathbf{s}}\rangle\langle\mu_{\mathbf{s}+\mathbf{r}}\rangle, \tag{7.3.5}$$

we deduce, from eqns (7.3.1) and (7.3.2), that, close to the critical point and for r much larger than L,

$$\begin{aligned} G(\tau, H, r) &= \langle\mu\rangle_L^2\{\langle\sigma_\alpha\sigma_{\alpha'}\rangle - \langle\sigma_\alpha\rangle\langle\sigma_{\alpha'}\rangle\} \\ &= \langle\mu\rangle_L^2 G(\tilde{\tau}, \tilde{H}, r/L) \end{aligned} \tag{7.3.6}$$

where we have assumed that the lattice vectors $\mathbf{s}$ and $\mathbf{s}+\mathbf{r}$ are in cells α and α', respectively. It remains now to find expressions for $\tilde{\tau}$ and $\tilde{H}$ in terms of τ and H.

An expression for $\tilde{H}$ is easily obtained by noting from definitions given in eqns (7.3.1) and (7.3.2) that the magnetic-field term in the iteration energy can be written as

$$H\sum_{\mathbf{r}}\mu_{\mathbf{r}} = HL^d\langle\mu\rangle_L\sum_\alpha\sigma_\alpha \tag{7.3.7}$$

from which we obtain

$$\tilde{H} = L^d\langle\mu\rangle_L H. \tag{7.3.8}$$

If we now assume that close to criticality $\langle\mu\rangle_L$, to leading order, is independent of τ and H and varies as L raised to some power, we have that

$$\tilde{H} \sim L^{da_H}H \qquad \text{as } \tau, H \to 0 \tag{7.3.9}$$

where

$$\langle\mu\rangle_L \sim L^{d(a_H-1)} \qquad \text{as } \tau, H \to 0. \tag{7.3.10}$$

Similarly, if we assume, to leading order, that

$$\tilde{\tau} \sim L^{da_\tau}|\tau| \qquad \text{as } \tau, H \to 0, \tag{7.3.11}$$

eqn (7.3.4) reduces to eqn (7.2.1) with the scaling parameter $\lambda = L^d$. Equations

(7.3.4), (7.3.9), and (7.3.11) are therefore equivalent to the scaling assumption of the previous section from which the standard scaling relations for critical exponents followed.

An added bonus of the block-spin picture is that from eqns (7.3.6) and (7.3.9)–(7.3.11) we can deduce a scaling form for the pair-correlation function. In particular, if we set the field H equal to zero, we obtain

$$G(\tau, r) \equiv G(\tau, 0, r) = L^{2d(a_H - 1)} G(\tau L^{da_\tau}, 0, r/L). \tag{7.3.12}$$

Assuming that this relation holds for r actually equal to L, we obtain on setting $L = r$ the scaling form

$$G(\tau, r) = r^{2d(a_H - 1)} g(\tau r^{da_\tau}) \tag{7.3.13}$$

where

$$g(x) = G(x, 0, 1). \tag{7.3.14}$$

If we now assume exponential decay of the correlation function of the form

$$G(\tau, r) \sim \exp(-r/\xi(\tau)) \qquad \tau \text{ fixed and } r \to \infty, \tag{7.3.15}$$

we deduce that the correlation-length $\xi(\tau)$ exponent ν defined by

$$\xi(\tau) \sim \tau^{-\nu} \qquad \text{as } \tau \to 0+ \tag{7.3.16}$$

is given from eqn (7.3.13) by

$$\nu = 1/da_\tau. \tag{7.3.17}$$

Similarly, if we assume as in Section 5.4 that the correlation function at the critical point $\tau = 0$ has power-law decay characterized by the exponent η defined by

$$G(0, r) \sim r^{-(d-2+\eta)}, \tag{7.3.18}$$

we deduce from eqn (7.3.13) that

$$2d(a_H - 1) = 2 - d - \eta. \tag{7.3.19}$$

In the previous section, we expressed the standard critical exponents α, β, γ, and δ in terms of a_τ and a_H. Specifically, from eqns (7.2.4) (noting that $\Delta = a_H/a_\tau$), we obtain

$$1/a_\tau = 2 - \alpha \tag{7.3.20}$$

and

$$2a_H = 1 + a_\tau \gamma. \tag{7.3.21}$$

Substituting these expressions into eqns (7.3.17) and (7.3.19) we easily obtain the additional scaling relations

$$d\nu = 2 - \alpha \tag{7.3.22}$$

and

$$\gamma = (2 - \eta)\nu. \tag{7.3.23}$$

Equation (7.3.23) will be recognized as the relation we deduced previously in Section 5.5 from the fluctuation relation. The so-called *hyperscaling* relation given in eqn (7.3.22) is new and has the interesting feature that it involves the dimensionality of the system. This relation is certainly valid for the two-dimensional Ising model ($\nu = 1, \alpha = 0_{\ln}$) but there is still some debate about its validity for the three-dimensional model. The classical values $\nu = 1/2$, $\alpha = 0_{\text{disc}}$ 'fit' eqn (7.3.22) in dimension $d = 4$ but it is almost certain that hyperscaling breaks down for $d > 4$ since it is known rigorously (Aizenman 1981) that, for $d > 4$, some critical exponents (e.g. γ) take on their classical values and that the Curie–Weiss theory holds rigorously in the limit $d \to \infty$ (Pearce and Thompson 1978).

7.4 The renormalization-group recipe

As we saw in the previous section, the problem in dealing with critical phenomena is how to handle systems with many degrees of freedom with correlations extending over many lattice spacings. The block-spin idea in effect introduces a transformation which preserves the structure of the system and related temperature and field variables by changing the length scale by a factor of $L > 1$ and at the same time reducing the number of degrees of freedom by a factor L^d. The block-spin approach results in scaling relations between critical exponents but, in the absence of any analytical form for the block-spin transformation, the method is incapable of providing a means for actually calculating critical exponents.

On reflection, it is probably asking too much of the block-spin method to expect it to yield a transformation that maps a nearest-neighbour interacting system into itself. Rather, one might expect such a transformation to change the form of the interactions while hopefully preserving the symmetries and dimensionality of the system. This is the basic idea behind the renormalization-group approach to critical phenomena first introduced by Wilson (1971).

Thus, instead of considering a particular Hamiltonian system, we consider a space of Hamiltonians and an operator $\mathbb{R}_L$ acting on this space, characterized by L and having the properties attributed to the block-spin transformation, namely, that, when $\mathbb{R}_L$ is applied to a particular Hamiltonian system, the correlation length is reduced by a factor L and the number of degrees of freedom is reduced by a factor L^d, while preserving the symmetries and dimensionality d of the system.

For convenience and simplicity, we will restrict our discussion here to transformations on real space, as opposed to momentum space which was treated originally by Wilson, and consider only magnetic systems whose Hamiltonians may be characterized by a set of coupling constants expressed as a vector

$$\mathbf{K} = (K^{(1)}, K^{(2)}, \ldots). \tag{7.4.1}$$

We will assume that the dimension of $\mathbf{K}$ is sufficiently large to include all possible interactions, from nearest-neighbour $K^{(1)}$, external field $K^{(2)}$, next-nearest neighbour, three-spin, four-spin, etc. that may be generated by applying $\mathbb{R}_L$ to an arbitrary Hamiltonian in the space.

The renormalization-group operator $\mathbb{R}_L$ transforms an arbitrary system in the space, with coupling constant vector $\mathbf{K}$, into another system in the space specified by eqn (7.4.1), with coupling constant vector $\mathbf{K}'$ say, given by

$$\mathbf{K}' = \mathbb{R}_L \circ \mathbf{K}. \tag{7.4.2}$$

Alternatively, eqn (7.4.2) may be expressed in terms of the components of $\mathbf{K}$ and $\mathbf{K}'$ by

$$K^{(i)\prime} = f_L^{(i)}(K^{(1)}, K^{(2)}, \ldots), \qquad i = 1,2, \ldots \tag{7.4.3}$$

and, in general, one can expect the operator $\mathbb{R}_L$, or equivalently the functions $f_L^{(i)}$, to be non-linear. Because of this non-linearity and the fact that $\mathbb{R}_L$ is generally constructed by manipulating the Boltzmann factor $\exp(-\beta\mathscr{H})$, it is important to include the multiplicative factor β in the definition of the coupling constants $K^{(i)}$ when initially applying $\mathbb{R}_L$ to a particular system. That is, the components of the initial $\mathbf{K}_0$ are the actual coupling constants multiplied by β, whereas the components of $\mathbf{K}' = \mathbb{R}_L \circ \mathbf{K}_0$ will, in general, depend non-linearly on β.

The construction of particular operators may be thought of as the first stage of the renormalization-group approach. The second stage, which is the more interesting and conceptually appealing part of the approach is what we refer to as the renormalization-group recipe. The recipe is reasonably precise but, unfortunately, with only a few exceptions, the construction of particular operators $\mathbb{R}_L$ is at best only approximate.

For the moment let us assume that we have an $\mathbb{R}_L$ with the properties described above. That is, the correlation length ξ and the singular part of the free energy ψ_s of two systems related by eqn (7.4.2) satisfy

$$\xi(\mathbf{K}') = L^{-1}\xi(\mathbf{K}) \qquad \text{and} \qquad \psi_s(\mathbf{K}') = L^d\psi_s(\mathbf{K}). \tag{7.4.4}$$

The renormalization-group recipe then proceeds as follows.

Starting from a particular system specified by $\mathbf{K}_0$ we repeatedly apply $\mathbb{R}_L$ to obtain a sequence of systems characterized by $\mathbf{K}_1, \mathbf{K}_2, \ldots$ defined recursively by

$$\mathbf{K}_{n+1} = \mathbb{R}_L \circ \mathbf{K}_n, \qquad n = 0, 1, 2, \ldots. \tag{7.4.5}$$

From eqn (7.4.4), we then have

$$\xi(\mathbf{K}_{n+1}) = L^{-1}\xi(\mathbf{K}_n) = \cdots = L^{-n}\xi(\mathbf{K}_0). \tag{7.4.6}$$

Suppose now that the sequence $\{\mathbf{K}_n\}$ approaches $\mathbf{K}^*$ as $n \to \infty$. Assuming continuity of $\mathbb{R}_L$ it follows that $\mathbf{K}^*$ is a *fixed point* of $\mathbb{R}_L$. That is, from eqn (7.4.5),

$$\mathbf{K}^* = \lim_{n\to\infty} \mathbf{K}_{n+1} = \mathbb{R}_L \circ \left(\lim_{n\to\infty} \mathbf{K}_n \right) = \mathbb{R}_L \circ \mathbf{K}^*. \tag{7.4.7}$$

So long as $\mathbf{K}^*$ is finite it corresponds to some Hamiltonian system in our space at a finite temperature. Moreover, from eqn (7.4.4) with $\mathbf{K} = \mathbf{K}^*$, we have $\xi(\mathbf{K}^*) = L^{-1}\xi(\mathbf{K}^*)$ so that $\xi(\mathbf{K}^*)$ must be either zero or infinite. If we exclude for the moment the possibility that $\xi(\mathbf{K}^*) = 0$, it follows, from eqn (7.4.6), that the correlation length of the initial system

$$\xi(\mathbf{K}_0) = \lim_{n\to\infty} L^n \xi(\mathbf{K}_{n+1}) \tag{7.4.8}$$

is also infinite, which means that we have commenced the iteration eqn (7.4.5) with our initial system at a critical point. For convenience, we will refer to fixed points $\mathbf{K}^*$ of systems with infinite correlation length as physical fixed points.

As an example, if we choose as our initial system a model with nearest-neighbour interactions only and in zero field, the initial vector has the form

$$\mathbf{K}_0 = (\beta J^{(1)}, 0, 0, \ldots) \tag{7.4.9}$$

where $J^{(1)}$ denotes the coupling constant between nearest neighbours. Assuming that our system eqn (7.4.9) has a critical point at $\beta = \beta_c$, the sequence $\{\mathbf{K}_n\}$ defined by eqn (7.4.5), and starting from $\mathbf{K}_0$, eqn (7.4.9), for the particular value $\beta = \beta_c$, will approach a physical fixed point of $\mathbb{R}_L$.

In general, if we start the iteration eqn (7.4.5) away from criticality, that is, from $\mathbf{K}_0$ with $\beta \neq \beta_c$ so that $\xi(\mathbf{K}_0)$ is finite, we have from eqn (7.4.8), since $L > 1$, that

$$\lim_{n\to\infty} \xi(\mathbf{K}_n) = 0 \qquad (\beta \neq \beta_c). \tag{7.4.10}$$

It follows that physical fixed points are unstable.

For $\beta > \beta_c$ one might expect the components of $\mathbf{K}_n$ to diverge to the 'low-temperature' fixed point while for $\beta < \beta_c$ one might expect $\mathbf{K}_n$ to approach the zero vector corresponding to the 'high-temperature' fixed point $\mathbf{K}^* = 0$. Both of these fixed points would then be stable and would correspond to limiting situations having zero correlation length in agreement with eqn (7.4.10).

The above are not the only possibilities of course. A zero correlation length could, for example, occur at a finite $\mathbf{K}$. Also, there is no reason to suppose that an $\mathbb{R}_L$ will have only fixed points. Iterates could, for example, approach strange attractors in parameter space, show periodic behaviour, or be totally chaotic. In general, however, one would expect an $\mathbb{R}_L$ to have at least one physical fixed point. Such fixed points are properties of $\mathbb{R}_L$ itself and not of some set of Hamiltonian systems. Each fixed point $\mathbf{K}^*$ of $\mathbb{R}_L$, however, will have a domain of attraction, or subspace of Hamiltonian systems, that approach $\mathbf{K}^*$ at criticality. Such subspaces will then constitute the so-called *universality classes* of Hamiltonian systems having a common set of critical exponents. The problem of universality is then equivalent to classifying the domains of attraction of physical fixed points of renormalization group transformations. In terms of parameter or $\mathbf{K}$-space, these domains of attraction are usually referred to as *critical surfaces* or, in more mathematical terms, as *stable manifolds*. Starting the iteration eqn (7.4.5) at a $\mathbf{K}_0$ with $\beta=\beta_c$ then means that we are on the critical surface of some physical fixed point. A simple example of this behaviour is given in the following section and illustrated in Fig. 7.3.

Suppose now that we start our iteration with $\mathbf{K}_0$ close to its critical value $\mathbf{K}_{0,c}$ by choosing β close to β_c. Although, as we have pointed out, the physical fixed point $\mathbf{K}^*$ approached at criticality is unstable, one would expect, for β sufficiently close to β_c, the iterates $\mathbf{K}_n$ to initially approach $\mathbf{K}^*$ for $n \sim N$ before subsequently diverging away from the critical surface. Assuming this to be the case, we can then write

$$\mathbf{K}_n = \mathbf{K}^* + \mathbf{k}_n \tag{7.4.11}$$

with the k_ns small for $n \gtrsim N$. Substituting eqn (7.4.11) into eqn (7.4.5) and expanding the right-hand side in the $\mathbf{k}_n$s we obtain, on neglecting high-order terms, the linear form

$$\mathbf{k}_{n+1} = \mathbb{A}_L(\mathbf{K}^*) \circ \mathbf{k}_n \tag{7.4.12}$$

of eqn (7.4.5) where $\mathbb{A}_L(\mathbf{K}^*)$ denotes the linearization of $\mathbb{R}_L$ around $\mathbf{K}^*$.

If we now assume that $\mathbb{A}_L(\mathbf{K}^*)$ has a complete set of eigenvectors $\boldsymbol{\phi}_i$ with corresponding eigenvalues Λ_i, $i=1, 2, 3, \ldots$, that is

$$\mathbb{A}_L(\mathbf{K}^*) \circ \boldsymbol{\phi}_i = \Lambda_i \boldsymbol{\phi}_i, \qquad i=1, 2, 3, \ldots, \tag{7.4.13}$$

we can construct the eigenvector expansion

$$\mathbf{k}_n = \sum_i u_i^{(n)} \boldsymbol{\phi}_i \tag{7.4.14}$$

and apply $\mathbb{A}_L(\mathbf{K}^*)$ n times from $n=N$ to obtain

$$\mathbf{k}_{n+N} = \mathbb{A}_L^n \circ \mathbf{k}_N = \sum_i u_i^{(N)} \Lambda_i^n \boldsymbol{\phi}_i \tag{7.4.15}$$

which gives, from eqn (7.4.14),

$$u_i^{(N+n)} = \Lambda_i^n u_i^{(N)}, \qquad n=0, 1, \ldots. \tag{7.4.16}$$

The influence of the so-called *scaling fields* $u_i^{(N)}$, which we will denote by u_i henceforth, will obviously grow and become *relevant* if their associated eigenvalue Λ_i exceeds unity in absolute value. Similarly, if $|\Lambda_i|<1$, the associated scaling field is *irrelevant*. The *marginal* case $|\Lambda_i|=1$ poses some interesting problems which we will not enter into here.

If we now parametrize our system *close to criticality* by the scaling fields u_i, which may be thought of as generalized coordinates, the correlation length from eqns (7.4.6) and (7.4.16) behaves as

$$\xi(u_1, u_2, \ldots) = L^n \xi(\Lambda_1^n u_1, \Lambda_2^n u_2, \ldots). \tag{7.4.17}$$

Similarly, from eqns (7.4.4) and (7.4.16), the free energy close to criticality satisfies

$$\psi_s(u_1, u_2, \ldots) = L^{-nd} \psi_s(\Lambda_1^n u_1, \Lambda_2^n u_2, \ldots). \tag{7.4.18}$$

If we are precisely at criticality when we commence the iteration (7.4.5), $\mathbf{K}_n \to \mathbf{K}^*$ and hence $\mathbf{k}_n \to \mathbf{0}$ as $n \to \infty$. It follows that the iterated scaling fields $u_i^{(n)}$ must approach zero at critically as $n \to \infty$. From eqn (7.4.16) this will always be the case for irrelevant scaling fields but for relevant fields we require u_i to vanish identically at criticality. Viewed another way one could say that, in the space described by the coordinates u_i, the origin corresponds to criticality and that, in the neighbourhood of this point on the subspace or *critical surface* for which the relevant scaling fields are set equal to zero, all points approach the origin under the action of $\mathbb{R}_L$. This is simply another expression of the notion of universality class.

So far as critical behaviour is concerned, we need only take account of the relevant scaling fields, which in most cases can be deduced on physical grounds. For example, if we consider a ferromagnetic system in zero field, the critical point is determined uniquely by setting the temperature T equal to its critical value T_c. Since relevant scaling fields must vanish at $T = T_c$, it follows that, in general for such systems, only one field can be relevant since otherwise one would have too many conditions on T_c. We therefore expect for such cases that

$$\Lambda_1 > 1 > \Lambda_2 \geqslant \Lambda_3 \geqslant \ldots. \tag{7.4.19}$$

Also if we assume, as is customary, that the $\mathbf{K}_n$s are regular functions of the temperature T, it follows that with $\tau = (T/T_c) - 1$, the only relevant scaling field has the form

$$u_1 = a\tau + O(\tau^2) \qquad \text{as } \tau \to 0. \tag{7.4.20}$$

Recalling now the definition $\xi(\tau) \sim \tau^{-\nu}$ as $\tau \to 0+$ of the correlation length critical exponent ν, it follows from eqns (7.4.17), (7.4.19), and (7.4.20) that

$$\xi(\tau) = L^n \xi(\Lambda_1^n \tau) \sim (L\Lambda_1^{-\nu})^n \tau^{-\nu} \qquad \text{as } \tau \to 0+ \tag{7.4.21}$$

and hence we require

$$L\Lambda_1^{-\nu} = 1 \qquad \text{or} \qquad \nu = \ln L / \ln \Lambda_1 \tag{7.4.22}$$

which gives an explicit expression for the exponent ν.

At first sight eqn (7.4.22) is somewhat perturbing since ν appears to depend on the length scale L. On physical grounds, however, we require successive renormalizations by factors L and L' to be equivalent to a single renormalization by a factor LL'. In terms of the renormalization-group transformation this means that

$$\mathbb{R}_{L'} \circ \mathbb{R}_L = \mathbb{R}_{LL'}. \tag{7.4.23}$$

This requirement actually makes the set of $\mathbb{R}$s a *semi-group*, rather than a group in the strict mathematical sense, and restricts the eigenvalues Λ_i to the form

$$\Lambda_i = L^{\mu_i} \tag{7.4.24}$$

so that from eqn (7.4.22) ν is, as it should be, independent of L.

In the case of non-zero field H we have a second relevant field u_2 which must vanish when $H=0$, so again, from the regularity assumption, u_2 has the form

$$u_2 = bH + O(H^2) \qquad \text{as } H \to 0. \tag{7.4.25}$$

If we now, from eqn (7.4.24), define the scaling exponents a_τ and a_H by

$$\Lambda_1 = L^{da_\tau} \qquad \text{and} \qquad \Lambda_2 = L^{da_H} \tag{7.4.26}$$

corresponding to scaling fields u_1 and u_2 given near criticality by eqns (7.4.20) and (7.4.25), respectively, the singular part of the free energy, from eqn (7.4.18), has the form

$$\begin{aligned}\psi_s(\tau, H) &= L^{-nd}\psi_s(L^{nda_\tau}\tau, L^{nda_H}H)\\ &= \lambda^{-1}\psi_s(\lambda^{a_\tau}\tau, \lambda^{a_H}H)\end{aligned} \tag{7.4.27}$$

where

$$\lambda = L^{dn}. \tag{7.4.28}$$

Equation (7.4.27) will be recognized immediately as the scaling assumption eqn (7.2.1) which we derived heuristically in the previous section from the Kadanoff block-spin picture.

In Section 7.2, we saw that critical exponents can be expressed in terms of a_τ and a_H. Unlike scaling theory, however, which only yields relations between critical exponents, the renormalization-group method provides a recipe for actually calculating a_τ and a_H in terms of the eigenvalues (7.4.26) of a certain operator obtained by linearizing a renormalization-group transformation around an appropriate fixed point. Values for critical exponents can then be obtained from scaling expressions. For example, from eqns (7.4.22) and (7.4.26),

$$\nu = (da_\tau)^{-1}. \tag{7.4.29}$$

Although, as we have remarked, the above recipe is reasonably precise it is not without its share of difficulties. We made several crucial assumptions in the above discussion, for example, the regularity assumption on $\mathbb{R}_L$ which amounts to assuming that the components of the coupling constants of the transformed $\mathbf{K}' = \mathbb{R}_L \circ \mathbf{K}$ are regular or analytic functions of the components of **K** and, in particular, of temperature. There is also the question of the existence of fixed points of $\mathbb{R}_L$ and what one actually means by convergence and continuity in the (effectively) infinite-dimensional space on which $\mathbb{R}_L$ acts. In addition, we have assumed that the linearized version of $\mathbb{R}_L$ around a fixed point has a complete set of eigenvectors and in the process of linearization we have used the notion of 'close to' in the Hamiltonian space which presupposes the existence of a norm or metric. All of the above could present severe mathematical difficulties in the execution of the renormalization-group recipe. Some of the potential pitfalls and problems with the method have in fact been highlighted and discussed in some detail by Griffiths and Pearce (1979).

7.5 Some examples of renormalization-group transformations

As mentioned in the previous section, the renormalization-group approach consists of two parts: (1) the construction of a renormalization-group transformation $\mathbb{R}_L$; and (2) a recipe to follow once $\mathbb{R}_L$ has been constructed. Many examples of renormalization-group transformations have appeared in the literature and, with only a few exceptions, they all contain uncontrolled approximations, so that there is no way of estimating errors in the final results unless of course one knows the exact result by using other methods. This is the main drawback of the method. In this section, we give two simple examples of $\mathbb{R}_L$s—one exact and one approximate. For reviews, other examples, and references to what has become an enormous literature on the subject the reader is referred to Domb and Green (1976), Barber (1977), and the written version of the Nobel Prize lecture given by Wilson (1983).

Perhaps the simplest example and one of the few exactly renormalizable models is the one-dimensional Ising model with nearest-neighbour interactions and energy given by

$$E\{\mu\} = -J \sum_{i=1}^{2N} \mu_i \mu_{i+1} \qquad (\mu_i = \pm 1,\ \mu_{2N+1} \equiv \mu_1). \tag{7.5.1}$$

The number of degrees of freedom ($2N$) can be reduced by a factor of two and the lattice spacing increased by a factor of two ($L=2$) by summing over the even labelled spins $\mu_2, \mu_4, \ldots, \mu_{2N}$ in the partition function

$$Z_{2N}(K) = \sum_{\{\mu\}} \exp\left(K \sum_{i=1}^{2N} \mu_i \mu_{i+1}\right), \qquad K = \beta J. \tag{7.5.2}$$

Using the fact that

$$\begin{aligned}\sum_{\mu=\pm 1} \exp[K\mu(\mu_1+\mu_2)] &= 2\cosh[K(\mu_1+\mu_2)] \\ &= 2(\cosh 2K)^{1/2} \exp(K'\mu_1\mu_2)\end{aligned} \tag{7.5.3}$$

where

$$K' = \tfrac{1}{2}\ln(\cosh 2K), \tag{7.5.4}$$

it is easily verified that

$$Z_{2N}(K)=2^N(\cosh 2K)^{N/2}Z_N(K'). \tag{7.5.5}$$

Equation (7.5.4) is the required renormalization-group transformation, yielding the sequence $\{K_n\}$ of coupling constants defined iteratively by

$$K_{n+1}=\mathbb{R}_2\circ K_n=\tfrac{1}{2}\ln(\cosh 2K_n), \qquad K_0=K. \tag{7.5.6}$$

Unfortunately, the transformation $\mathbb{R}_2$ defined in eqn (7.5.6) has no non-trivial fixed points, which is as it should be, of course, since the model has no phase transition. So while the transformation is exact, it is rather uninteresting since the recipe developed in the previous section cannot be implemented.

The process described above of selectively summing over subsets of spins, called *spin decimation*, can be repeated for the Ising model on a square lattice by summing over spins on one sublattice, for example, the sites marked by crosses in Fig. 7.1.

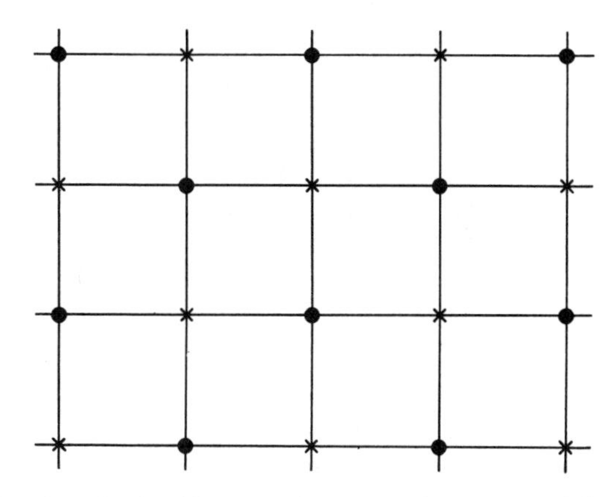

FIG. 7.1 Decimation obtained by summing over spins on one of the two sublattices.

If we focus our attention on one spin μ in a sublattice and its four nearest-neighbour spins $\mu_1, \mu_2, \mu_3, \mu_4$ on the other sublattice and perform the sum over $\mu=\pm 1$ we induce the so-called *star-square transformation* shown in Fig. 7.2. This transformation induces nearest-neighbour, next-nearest-neighbour, and four-spin interactions, increases the lattice spacing by a factor of $L=\sqrt{2}$ and decreases the number of degrees of freedom by $L^d=2$.

A simple calculation gives

$$\sum_{\mu=\pm 1}\exp[K\mu(\mu_1+\mu_2+\mu_3+\mu_4)]=2\cosh[K(\mu_1+\mu_2+\mu_3+\mu_4)]$$

$$=A\exp\left[K'\sum_{1\leq i<j\leq 4}\mu_i\mu_j+U\mu_1\mu_2\mu_3\mu_4\right] \tag{7.5.7}$$

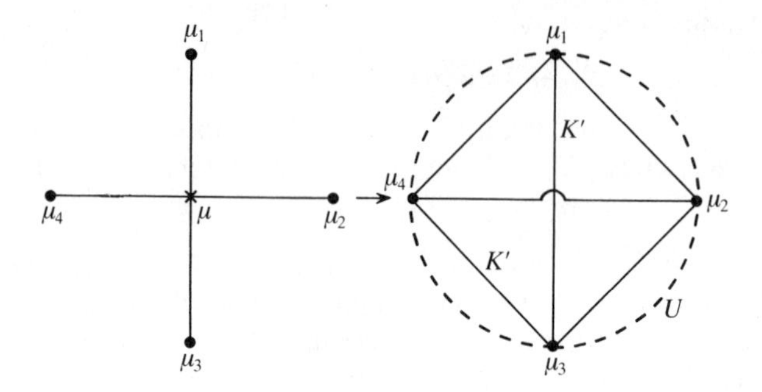

FIG. 7.2 The star-square transformation.

where

$$A = 2(\cosh^4 2K \cosh 4K)^{1/8}, \tag{7.5.8}$$

$$K' = \tfrac{1}{8} \ln (\cosh 4K), \tag{7.5.9}$$

and

$$U = \tfrac{1}{8} \ln (1 - \tanh^4 2K). \tag{7.5.10}$$

Equations (7.5.9) and (7.5.10) specify the action of the decimation transformation on the *particular* Hamiltonian system consisting of nearest-neighbour interactions only. The new nearest-neighbour interaction K_1 is in fact equal to $2K'$ since the new bond is shared by two spins on the original sublattice. The new next-nearest-neighbour coupling constant L_1 is equal to K' (and originally zero) and the new four-spin coupling constant is U.

Unfortunately, it is extremely difficult to repeat the decimation transformation beyond the first few steps since after each application of the transformation more and more complicated couplings are introduced. It is thus virtually impossible to construct the required renormalization-group transformation on the full space of coupling constants **K** described in the previous section. Wilson (1975), however, managed the construction with up to 200 coupling constants! Here, following Wilson, we retain only two in an *approximate* renormalization-group transformation.

The approximation begins by assuming that K is small, corresponding to high temperatures. From eqn (7.5.10), U is then very small so we set it equal to zero. The nearest- and next-nearest-neighbour coupling constants after one application of the decimation transformation are then given approximately from eqn (7.5.9) by

$$K_1 = 2K^2 \qquad \text{and} \qquad L_1 = K^2, \tag{7.5.11}$$

respectively. If we now introduce a small next-nearest-neighbour coupling constant L_0 initially and treat it as a perturbation we arrive after further approximation at the following renormalization group equations

$$\left.\begin{aligned} K_{n+1} &= 2K_n^2 + L_n \\ L_{n+1} &= K_n^2 \end{aligned}\right\}, \qquad n=0, 1, \ldots \tag{7.5.12}$$

where $K_0 = K$.

Following the renormalization-group recipe we first find the non-trivial fixed points of eqn (7.5.12). It is easily established that there is only one with

$$K_n = K^* = 1/3 \qquad \text{and} \qquad L_n = L^* = 1/9. \tag{7.5.13}$$

Next, to find the critical point of the nearest-neighbour model we set $L_0 = 0$, vary the initial $K_0 = K$, and iterate eqn (7.5.12) until we find the *critical* K_c for which $K_n \to K^*$ and $L_n \to L^*$ as $n \to \infty$. The value one finds for K_c is

$$K_c = 0.3921 \ldots \tag{7.5.14}$$

which is to be compared with Onsager's exact value of

$$K_c = 0.4407 \ldots . \tag{7.5.15}$$

Finally, to determine approximate values for critical exponents we linearize eqn (7.5.12) about the fixed point eqn (7.5.13). That is, we set

$$K_n = K^* + x_n, \qquad L_n = L^* + y_n \tag{7.5.16}$$

in eqn (7.5.12), expand the right-hand sides, and retain only linear terms in x_n and y_n to obtain the linearized equations

$$\left.\begin{aligned} x_{n+1} &= \tfrac{4}{3}x_n + y_n \\ y_{n+1} &= \tfrac{2}{3}x_n \end{aligned}\right\}, \qquad . \; n=0, 1, 2, \ldots \; . \tag{7.5.17}$$

The approximate linearized decimation transformation is then the 2×2 matrix

$$\mathbb{A} = \begin{pmatrix} \frac{4}{3} & 1 \\ \frac{2}{3} & 0 \end{pmatrix} \tag{7.5.18}$$

with eigenvalues

$$\Lambda_1 = \tfrac{1}{3}(2 + \sqrt{10}) \qquad \text{and} \qquad \Lambda_2 = \tfrac{1}{3}(2 - \sqrt{10}). \tag{7.5.19}$$

As expected Λ_1 is the only relevant eigenvalue and from eqn (7.4.22), for example, we obtain the critical exponent

$$\nu = \ln \sqrt{2} / \ln \Lambda_1 = 0.638 \ldots \tag{7.5.20}$$

which is to be compared with the exact value $\nu = 1$.

Universality is also incorporated in the above approximation when one considers the case of $L_0 \neq 0$. Thus, since eqn (7.5.12) has only one fixed point, it follows that for each L_0 there exists $K_0 = K_c(L_0)$ on the critical surface, as shown in Fig. 7.3, such that iterates of eqn (7.5.12) approach the fixed point. Since critical exponents are determined by linearizing around the fixed point,

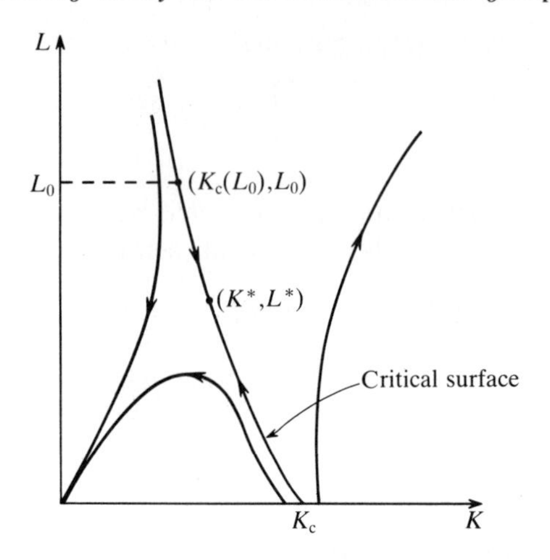

FIG. 7.3 Critical surface and fixed point of approximate decimation transformation.

which is independent of L_0, it then follows that the next-nearest-neighbour model and the nearest-neighbour model have the same critical exponents.

We also observe from Fig. 7.3 that, when the iteration is commenced at some point off the critical surface, the iterates either diverge to the low-temperature fixed point or converge to the high-temperature fixed point, as discussed in the previous section.

7.6 Problems

1. If $f(\lambda x)=g(\lambda)f(x)$ for all real λ and x and g is differentiable, prove that $g(\lambda)=\lambda^p$ for some real p.

2. From the scaling assumption (7.2.1) show that the singular part of the isothermal susceptibility has the scaling form

$$\chi(\tau, H)=\lambda^{\gamma/d\nu}\chi(\tau\lambda^{a_\tau}, H\lambda^{a_H}).$$

3. From the scaling assumption (7.2.2) show that the reduced magnetization $|\tau|^{-\beta}m(\tau, H)$ is a function only of the reduced field $H|\tau|^{-\beta\delta}$.

4. Show that the scaling assumption (7.3.12) for the pair-correlation function $G(\tau, r)$ is equivalent to the scaling form

$$G(\tau, r)=\xi^{2-d-\eta}f(r/\xi)$$

where ξ is the correlation length.

5. Use the method of decimation to show that the zero-field free energy of the one-dimensional spherical model with nearest-neighbour coupling constant $J=K/\beta$ is

given by

$$-\beta\psi(\beta,0)=\frac{1}{2}\left\{(1+4K^2)^{1/2}-\ln\frac{1}{4\pi}[1+(1+4K^2)^{1/2}]\right\}$$

(cf. Problem 10 of Chapter 6).

6. Wilson's approximate renormalization-group equations (Wilson 1971) for d dimensions and a length scale factor of $L=2$ are

$$r_{l+1}=4\left(r_l+\frac{3u_l}{1+r_l}-9u_l^2\right)$$

and

$$u_{l+1}=2^{4-d}(u_l-9u_l^2).$$

Deduce that, when $d>4$, exponents are classical and that, when $\varepsilon=4-d>0$ is 'small',

$$\nu=\frac{1}{2}+\frac{\varepsilon}{12}+O(\varepsilon^2).$$

(This is the so-called ε-expansion of Wilson and Fisher 1972.)

7. The semi-group property eqn (7.4.23) states that a renormalization-group transformation $\mathbb{R}_\lambda$ on a one-dimensional Hamiltonian space with length scale factor λ satisfies

$$\mathbb{R}_{\lambda\mu}(x)=\mathbb{R}_\lambda(\mathbb{R}_\mu(x))=\mathbb{R}_\mu(\mathbb{R}_\lambda(x))$$

for all real λ and μ and $\mathbb{R}_1(x)=x$.

Assuming that $\mathbb{R}_\lambda$ is differentiable and that for each λ there exists a unique non-trivial fixed point x_λ *of* $\mathbb{R}_\lambda$, prove that $x_\lambda=x_c$ is independent of λ and that

$$\mathbb{R}'_\lambda(x_c)=\lambda^p$$

for some real p.

8. The free energy $\psi(x)$ of a system transforms under a renormalization-group transformation $\mathbb{R}_\lambda$ of the type discussed in Problem 7 as

$$\psi(x)=g(x)+\lambda^{-d}\psi(\mathbb{R}_\lambda(x)).$$

Assuming that $\psi(x)$ is singular and $g(x)$ is regular at x_c and that $\mathbb{R}'_\lambda(x_c)=\lambda^p$, show that the specific-heat exponent α, defined by

$$\psi_s(x)\sim A|x-x_c|^{2-\alpha} \qquad \text{as } x\to x_c,$$

is given by

$$\alpha=2-d/p.$$

Under what conditions would one expect

$$\psi(x)=\sum_{n=0}^{\infty}\lambda^{-nd}g(\mathbb{R}_\lambda^{(n)}(x))$$

where $\mathbb{R}_\lambda^{(n)}$ is defined recursively by

$$\mathbb{R}_\lambda^{(n+1)}(x)=\mathbb{R}_\lambda(\mathbb{R}_\lambda^{(n)}(x))$$

with

$$\mathbb{R}_\lambda^{(1)}(x)=\mathbb{R}_\lambda(x)?$$

9. Obtain the most general renormalization-group transformation of the form

$$\mathbb{R}_\lambda(x)=a(\lambda)x[1+b(\lambda)x]^{-1}$$

and discuss the nature of its fixed points when $\lambda>1$.

7.7 Solutions to problems

1. Assuming $f(\lambda x)=g(\lambda)f(x)$ for all real λ and x, we have

$$\begin{aligned} f((\lambda\mu)x) &= g(\lambda\mu)f(x) \\ &= f(\lambda(\mu x)) \\ &= g(\lambda)f(\mu x) \\ &= g(\lambda)g(\mu)f(x) \end{aligned}$$

for all real λ, μ, and x. It follows that

$$g(\lambda\mu)=g(\lambda)g(\mu) \tag{1}$$

for all real λ and μ.

Differentiating eqn (1) with respect to μ and setting μ equal to unity, we obtain

$$\lambda g'(\lambda)=g(\lambda)g'(1)$$

from which the required result

$$g(\lambda)=\lambda^p$$

follows immediately with $p=g'(1)$.

2. Differentiating the scaling assumption eqn (7.2.1) twice with respect to H gives

$$\lambda\chi(\tau, H)=\lambda^{2a_H}\chi(\tau\lambda^{a_\tau}, H\lambda^{a_H}).$$

The required result follows by noting, from eqns (7.2.4) and (7.3.11), that

$$2a_H-1=\gamma a_\tau=\gamma/dv.$$

3. Differentiating the scaling assumption (7.2.2) once with respect to H gives

$$m(\tau, H)=|\tau|^{a_\tau^{-1}-\Delta}Q'_\pm(H|\tau|^{-\Delta}).$$

The required result follows from eqns (7.2.4), (7.2.9), and (7.2.10) by noting that

$$\frac{1}{a_\tau}=\Delta=\beta \qquad \text{and} \qquad \Delta=2-(\alpha+\beta)=\beta\delta.$$

4. The required result follows immediately from eqns (7.3.16), (7.3.17), and (7.3.19) by setting $L=\xi$ in eqn (7.3.13).

5. The partition function for the one-dimensional spherical model with $2N$ spins is given by

$$Z_{2N}(K, s)=\int_{-\infty}^{\infty}\cdots\int \exp\left(K\sum_{i=1}^{2N} x_i x_{i+1}-s\sum_{i=1}^{2N} x_i^2\right)\mathrm{d}x_1\cdots\mathrm{d}x_{2N}$$

where s is chosen so that

$$\left\langle\sum_{i=1}^{2N} x_i^2\right\rangle=-\frac{\partial}{\partial s}[\ln Z_{2N}(K, s)]=2N. \tag{1}$$

Integrating over even-numbered spins using

$$\int_{-\infty}^{\infty}\exp[-sx^2+Kx(x_1+x_3)]\,\mathrm{d}x$$
$$=(\pi/s)^{1/2}\exp[K^2x_1x_3/2s+K^2(x_1^2+x_3^2)/4s]$$

gives, with $y_i=x_{2i-1}$, $i=1, 2, \ldots, N$,

$$Z_{2N}(K, s)=\left(\frac{\pi}{s}\right)^{N/2}\int_{-\infty}^{\infty}\cdots\int\exp\left[\frac{K^2}{2s}\sum_{i=1}^{N} y_i y_{i+1}-\left(s-\frac{K^2}{2s}\right)\sum_{i=1}^{N} y_i^2\right]\mathrm{d}y_1\ldots\mathrm{d}y_N.$$

Changing variables to $x_i=(K/2s)^{1/2}y_i$ then gives

$$Z_{2N}(K, s)=(2\pi/K)^{N/2}Z_N(K, s') \tag{2}$$

where

$$s'=\frac{2s^2}{K}-K \tag{3}$$

may be interpreted as a renormalization-group transformation.

Defining

$$f_n(K, s)=2^{-n}\ln Z_{2^n}(K, s)$$

and iterating eqn (2) gives

$$f_{n+1}(K, s)=(2^{-1}-2^{-n-1})\ln(2\pi/K)+2^{-n}f_1(K, s_n) \tag{4}$$

where

$$f_1(K, s_n)=2^{-1}\ln[\pi(s_n^2-K^2)^{-1/2}] \tag{5}$$

and

$$s_{n+1}=2s_n^2/K-K \tag{6}$$

with $s_0=s$.

The solution of the recursion relation (6) is

$$s_n=K\cosh(2^n\theta_0)$$

where

$$\theta_0=\operatorname{arcosh}\left(\frac{s}{K}\right)=\ln\left[\frac{s}{K}+\left(\frac{s^2}{K^2}-1\right)^{1/2}\right].$$

Combining the above results then gives

$$f(K, s) = \lim_{n\to\infty} f_n(K,s)$$

$$= \frac{1}{2}\ln(2\pi/K) - \lim_{n\to\infty} 2^{-n-1}\ln[\sinh 2^n\theta_0]$$

$$= \frac{1}{2}[\ln(2\pi/K) - \theta_0]$$

$$= -\frac{1}{2}\ln\frac{1}{2\pi}[s+(s^2-K^2)^{1/2}].$$

The required result follows by noting that the spherical-model free energy ψ is given by

$$-\beta\psi = f(K, s) + s$$

where, from eqn (1), s is the solution of

$$\frac{\partial}{\partial s}[f(K, s)+s] = 0,$$

that is

$$s = \tfrac{1}{2}(1+4K^2)^{1/2}.$$

6. In a certain sense the Wilson scheme corresponds to a modified Ising or Gaussian model with spin distribution

$$p_l(x) \sim \exp(-r_l x^2 - u_l x^4)$$

at the lth stage of the renormalization.

Initially, one must choose $u_0 \geqslant 0$ to avoid divergent integrals and then vary r_0 to locate the 'critical point' $r_c(u_0)$. From the initial point $(r_c(u_0), u_0)$, the renormalization scheme then iterates along the 'critical surface' to the appropriate physical fixed point.

In general, the Wilson renormalization-group equations have two fixed points; the trivial or Gaussian fixed point $r=u=0$ and a non-trivial fixed point which for small $\varepsilon = 4-d$ is given by

$$r^* = -\frac{4\varepsilon}{9}\ln\varepsilon + O(\varepsilon^2)$$

and

$$u^* = \frac{\varepsilon}{9}\ln 2 + O(\varepsilon^2).$$

When $d>4$, it is clear that $u_l \to 0$ so that $r=u=0$ is the physical fixed point. In this case $r_c = 0$ and the linearized renormalization-group operator is the

$$\begin{pmatrix} 4 & 0 \\ 0 & 2^{4-d} \end{pmatrix}.$$

The maximum eigenvalue is $\lambda_1 = 4$ and the exponent

$$\nu = \ln 2/\ln\lambda_1 = 1/2$$

takes its classical value.

When $\varepsilon = 4 - d$ is small and positive, (r^*, u^*) is the physical fixed point and $r_c(u_0) = -4u_0 + O(\varepsilon^2)$ (assuming r_0 and u_0 are of order ε).

The linearized renormalization-group operator in this case is given, to order ε, by

$$\begin{pmatrix} 4-12u^* & 12 \\ 0 & 1-9u^* \end{pmatrix}.$$

The maximum eigenvalue is $\lambda_1 = 4 - 12u^*$ and, to order ε,

$$\nu = \ln 2/\ln \lambda_1 = \frac{1}{2} + \frac{\varepsilon}{12} + O(\varepsilon^2).$$

7. Setting $x = x_\lambda$ in the semi-group relation and, using the fact that $\mathbb{R}_\lambda(x_\lambda) = x_\lambda$, we obtain

$$\mathbb{R}_\lambda(\mathbb{R}_\mu(x_\lambda)) = \mathbb{R}_\mu(x_\lambda)$$

which implies

$$\mathbb{R}_\mu(x_\lambda) = x_\lambda$$

and hence that $x_\lambda = x_\mu = x_c$ for all λ and μ.

Differentiating the semi-group property with respect to x and setting $x = x_c$, we obtain

$$\begin{aligned} f(\lambda\mu) &\equiv \mathbb{R}'_{\lambda\mu}(x_c) \\ &= \mathbb{R}'_\lambda(\mathbb{R}_\mu(x_c))\,\mathbb{R}'_\mu(x_c) \\ &= \mathbb{R}'_\lambda(x_c)\,\mathbb{R}'_\mu(x_c) \\ &= f(\lambda)f(\mu). \end{aligned}$$

The required result follows from an argument similar to the one given in Problem 1.

8. Expanding about x_c, we have

$$\begin{aligned} \mathbb{R}_\lambda(x) &= \mathbb{R}_\lambda(x_c) + (x - x_c)\,\mathbb{R}'_\lambda(x_c) + \cdots \\ &= x_c + (x - x_c)\lambda^p + \cdots. \end{aligned}$$

Comparing singular parts on both sides of the equation for ψ, we obtain

$$A|x - x_c|^{2-\alpha} \sim \lambda^{-d} A|(x - x_c)\lambda^p|^{2-\alpha} \quad \text{as } x \to x_c.$$

It follows that

$$\lambda^{-d}\lambda^{p(2-\alpha)} = 1$$

and hence that

$$\alpha = 2 - d/p.$$

(Notice that in standard terminology $\nu = 1/p$ is the correlation-length critical exponent.)

Iterating the equation for $\psi(x)$ N times, we have

$$\psi(x) = \sum_{n=0}^{N-1} \lambda^{-nd} g(\mathbb{R}_\lambda^{(n)}(x)) + \lambda^{-Nd}\psi(\mathbb{R}_\lambda^{(N)}(x)).$$

Obviously, the stated result holds provided

$$\lim_{N\to\infty} \lambda^{-Nd}\psi(\mathbb{R}_\lambda^{(N)}(x)) = 0.$$

On physical grounds one expects $R_\lambda^{(N)}(x)$ to approach either zero or infinity (the high- and low-temperature fixed points) when $x \neq x_c$ as $N \to \infty$. Provided ψ is reasonably-well behaved in these two limiting cases, the infinite-series expression for ψ should hold when the length scale factor λ exceeds unity.

9. The semi-group property requires

$$\begin{aligned}\mathbb{R}_\lambda(\mathbb{R}_\mu(x)) &= a(\lambda)a(\mu)x\{1+[b(\mu)+a(\mu)b(\lambda)]x\}^{-1}\\ &= a(\lambda\mu)x[1+b(\lambda\mu)x]^{-1}\end{aligned}$$

which gives

$$a(\lambda\mu)=a(\lambda)a(\mu)$$

and, by symmetry,

$$b(\lambda\mu)=b(\mu)+a(\mu)b(\lambda)=b(\lambda)+a(\lambda)b(\mu).$$

The first relation (see Problem 1) gives

$$a(\lambda)=\lambda^p$$

and, on rearranging the second relation, one obtains

$$b(\mu)[a(\mu)-1]^{-1}=b(\lambda)[a(\lambda)-1]^{-1}=\text{constant}.$$

The most general form of the transformation is then

$$\mathbb{R}_\lambda(x)=\lambda^p x\{1+c(\lambda^p-1)x\}^{-1}$$

where p and c are constants.

The fixed points of $\mathbb{R}_\lambda(x)$ are $x=0$, $1/c$. An elementary calculation shows that

$$\mathbb{R}'(0)=\lambda^p \qquad \text{and} \qquad \mathbb{R}'\left(\frac{1}{c}\right)=\lambda^{-p}$$

so that, when $\lambda>1$, the origin is unstable and the fixed point $x=1/c$ is stable.

These results should be compared with the more general results stated in Problem 7.

REFERENCES

Abraham, D.B. (1972). High temperature susceptibility bounds for the two-dimensional Ising model. *Phys. Lett.* **39A**, 357.

Abraham, D.B. (1973). Susceptibility and fluctuations in the Ising ferromagnet. *Phys. Lett.* **43A**, 163.

Aizenman, M. (1981). Proof of the triviality of ϕ_d^4 field theory and some related mean-field features of Ising models for $d > 4$. *Phys. Rev. Lett.* **47**, 1.

Alder, B.J. and Wainwright, T.E. (1962). Phase transition in elastic discs. *Phys. Rev.* **127**, 359.

Ashkin, J. and Teller, E. (1943). Statistics of two-dimensional lattices with four components. *Phys. Rev.* **64**, 178.

Baker, G.A. (1961). One-dimensional order–disorder model which approaches a phase transition. *Phys. Rev.* **122**, 1477.

Barber, M.N. (1977). An introduction to the fundamentals of the renormalization group in critical phenomena. *Phys. Rep.* **29C**, 1.

Barber, M.N. and Baxter, R.J. (1973). On the spontaneous order of the eight-vertex model. *J. Phys.* **C6**, 2913.

Barker, J.A. and Henderson, D. (1976). What is 'liquid'? Understanding the states of matter. *Rev. mod. Phys.* **48**, 587.

Barouch, E., McCoy, B.M., and Wu, T.T. (1973). Zero-field susceptibility of the two-dimensional Ising model near T_c. *Phys. Rev. Lett.* **31**, 1409.

Baxter, R.J. (1971). Eight-vertex model in lattice statistics. *Phys. Rev. Lett.* **26**, 832.

Baxter, R.J. (1972). Partition function of the eight-vertex lattice model. *Ann. Phys.* **70**, 193.

Baxter, R.J. (1973). Asymptotically degenerate maximum eigenvalue of the eight-vertex model transfer matrix and interfacial tension. *J. stat. Phys.* **8**, 25.

Baxter, R.J. (1980). Hard hexagons: Exact solution. *J. Phys.* **A13**, L61.

Baxter, R.J. (1981). Corner transfer matrices. *Physica* **106A**, 18.

Baxter, R.J. (1982). *Exactly solved models in statistical mechanics.* Academic Press, London.

Baxter, R.J. and Enting, I.G. (1978). 399th solution of the Ising model. *J. Phys.* **A11**, 2463.

Berlin, T.H. and Kac, M. (1952). The spherical model of a ferromagnet. *Phys. Rev.* **86**, 821.

Bernal, J.D. and Fowler, R.H. (1933). A theory of water and ionic solution, with particular reference to hydrogen and hydroxyl ions. *J. Chem. Phys.* **1**, 515.

Bethe, H.A. (1935). Statistical theory of superlattices. *Proc. R. Soc., London* **A150**, 552.

Bragg, W.L. and Williams, E.J. (1934). Effect of thermal agitation on atomic arrangement in alloys I. *Proc. R. Soc., London* **A145**, 699.

Bragg, W.L. and Williams, E.J. (1935). Effect of thermal agitation on atomic arrangement in alloys II. *Proc. R. Soc., London* **A151**, 540.

Brown, S.C. (1950). The caloric theory of heat. *Am. J. Phys.* **18**, 319.

Buchdahl, H.A. (1966). *The concepts of classical thermodynamics.* Cambridge University Press, New York.

Callen, H.B. (1960). *Thermodynamics.* Wiley, New York.

Carathéodory, C. (1909). Untersuchungen über die Grundlagen der Thermodynamik. *Math. Ann.* **67**, 355.

Carnot, S. (1824). *Reflexions sur la puissance motrice du feu et sur les machines propres à développer cette puissance.* Bachelier, Paris. [English translation in *Reflections on the motive power of fire . . .*, (ed. E. Mendoza). Dover, New York 1960.]

Clausius, R. (1850). Über die bewegende Kraft der Wärme und die Gesetze die sich daraus für die Wärmelehre selbst ableiten lassen. *Ann. Phys.* **79**, 368, 500. [English translation: *Phil. Mag.* **2**, 1 (1851). Also *Reflections on the motive power of fire . . .*, (ed. E. Mendoza). Dover, New York 1960.]

Davy, H. (1799). Essay on heat, light and the combinations of light. In *Contributions to Physical and Medical Knowledge Primarily from the West of England.* Collected by T. Beddoes, Bristol, 1799.

Domb, C. and Green, M.S. (1976). *Phase transitions and critical phenomena* (ed. C. Domb and M.S. Green), Vol. 6. Academic Press, London.

Domb, C. and Sykes, M.F. (1957). On the susceptibility of a ferromagnet above the Curie point. *Proc. R. Soc., London* **A240**, 214.

Essam, J.W. and Fisher, M.E. (1963). Padé approximant studies of the lattice gas and Ising ferromagnet below the critical point. *J. Chem. Phys.* **38**, 802.

Fisher, M.E. (1959). The susceptibility of the plane Ising model. *Physica* **25**, 521.

Fisher, M.E. (1962). On the theory of critical point density fluctuations. *Physica* **28**, 172.

Fisher, M.E. (1964*a*). The free energy of a macroscopic system. *Arch. rat. mech. Anal.* **17**, 377.

Fisher, M.E. (1964*b*). Correlation functions and the critical region of simple fluids. *J. math. Phys.* **5**, 944.

Fisher, M.E. (1967). The theory of equilibrium critical phenomena. *Rep. Progr. Phys.* **30**, 615.

Fisher, M.E. (1969). Rigorous inequalities for critical-point correlation exponents. *Phys. Rev.* **180**, 594.

Fisher, M.E. and Burford, R.J. (1967). Theory of critical-point scattering and correlations I: The Ising model. *Phys. Rev.* **156**, 583.

Frisch, H.L., Rivier, N., and Wyler, D. (1985). Classical hard-sphere fluid in infinitely many dimensions. *Phys. Rev. Lett.* **54**, 2061.

Gantmacher, F.R. (1964). *The theory of matrices*, Vol. 2. Chelsea, New York.

Gates, D.J. (1970). Correlation functions for classical systems in the van der Waals limit. *J. Phys.* **A3**, L11.

Ginibre, J. (1967). Rigorous lower bound on the compressibility of a classical system. *Phys. Lett.* **24A**, 223.

Ginibre, J. (1970). General formulation of Griffiths' inequalities. *Commun. math. Phys.* **16**, 310.

Glasser, M.L. (1970). Exact partition function for the two-dimensional Ising model. *Am. J. Phys.* **38**, 1033.

Grad, H. (1952). Statistical mechanics, thermodynamics, and fluid dynamics of systems with an arbitrary number of integrals. *Commun. pure appl. Math.* **5**, 455.

Griffiths, R.B. (1964). Peierls proof of spontaneous magnetization in a two-dimensional Ising ferromagnet. *Phys. Rev.* **136A**, 437.

Griffiths, R.B. (1965*a*). Thermodynamic inequality near the critical point for ferromagnets and fluids. *Phys. Rev. Lett.* **14**, 623.

Griffiths, R.B. (1965*b*). Microcanonical ensemble in quantum statistical mechanics. *J. math. Phys.* **6**, 1447.

Griffiths, R.B. (1967). Correlations in Ising ferromagnets I and II. *J. math. Phys.* **8**, 478, 484.

Griffiths, R.B. (1972). Rigorous results and theorems. In *Phase transitions and critical phenomena* (ed. C. Domb and M.S. Green), Vol. I, 59. Academic Press, London.

Griffiths, R.B. and Pearce, P.A. (1979). Mathematical properties of position-space renormalization group transformations. *J. stat. Phys.* **20**, 499.

Guggenheim, E.A. (1945). The principle of corresponding states. *J. Chem. Phys.* **13**, 253.

Hardy, G.H., Littlewood, J.E., and Pólya, G. (1964). *Inequalities.* Cambridge University Press, New York.

Heller, P. (1967). Experimental investigations of critical phenomena. *Rep. Progr. Phys.* **30**, 731.

Hemmer, P.C., Kac, M., and Uhlenbeck, G.E. (1964). On the van der Waals theory of the vapour–liquid equilibrium III: Discussion of the critical region. *J. math. Phys.* **5**, 60.

Huang, K. (1963). *Statistical mechanics.* Wiley, New York.

Ising, E. (1925). Beitrag zur Theorie des Ferromagnetismus. *Z. Phys.* **31**, 253.

Josephson, B.D. (1967). Inequality for the specific heat II: Application to critical phenomena. *Proc. phys. Soc.* **92**, 269.

Joule, J.P. (1845). On the existence of an equivalent relation between heat and the ordinary forms of mechanical power. *Phil. Mag.* **27**, 205.

Joyce, G.S. (1972). Critical properties of the spherical model. In *Phase transitions and critical phenomena* (ed. C. Domb and M.S. Green), Vol. 2. Academic Press, New York.

Kac, M. (1964). The work of T.H. Berlin in statistical mechanics—a personal reminiscence. *Phys. Today* **17**, 40.

Kac, M. and Helfand, E. (1963). Study of several lattice systems with long-range forces. *J. math. Phys.* **4**, 1078.

Kac, M. and Thompson, C.J. (1971). Spherical model and the infinite spin dimensionality limit. *Physica Norvegica* **5**, 163.

Kac, M. and Thompson, C.J. (1977). Correlation functions in the spherical and mean spherical models. *J. math. Phys.* **18**, 1650.

Kac, M., Uhlenbeck, G.E., and Hemmer, P.C. (1963). On the van der Waals theory of the vapour–liquid equilibrium I: Discussion of a one-dimensional model. *J. math. Phys.* **4**, 216.

Kac, M. and Ward, J.C. (1952). A combinatorial solution of the two-dimensional Ising model. *Phys. Rev.* **88**, 1332.

Kadanoff, L.P. (1966). Scaling laws for Ising models near T_c. *Physics* **2**, 263.

Kaufman, B. (1949). Crystal statistics II: Partition function evaluated by spinor analysis. *Phys. Rev.* **76**, 1232.

Kaufman, B. and Onsager, L. (1949). Crystal statistics III: Short-range order in a binary Ising lattice. *Phys. Rev.* **76**, 1244.

Khinchin, A.I. (1949). *Mathematical foundations of statistical mechanics.* Dover, New York.

Kramers, H.A. and Wannier, G.H. (1941). Statistics of the two-dimensional ferromagnet, I and II. *Phys. Rev.* **60**, 252, 263.

Lebowitz, J.L. and Penrose, O. (1966). Rigorous treatment of the van der Waals–Maxwell theory of the liquid–vapour transition. *J. math. Phys.* **7**, 98.

Lee, T.D. and Yang, C.N. (1952). Statistical theory of equation of state and phase transitions II: Lattice gas and Ising model. *Phys. Rev.* **87**, 410.

Lewis, H.W. and Wannier, G.H. (1952). Spherical model of a ferromagnet. *Phys. Rev.* **88**, 682.

Lieb, E.H. (1967*a*). Exact solution of the problem of the entropy of two-dimensional ice. *Phys. Rev. Lett.* **18**, 692.

Lieb, E.H. (1967*b*). Exact solution of the F model of an antiferroelectric. *Phys. Rev. Lett.* **18**, 1046.

Lieb, E.H. (1967*c*). Exact solution of the two-dimensional Slater KDP model of a ferroelectric. *Phys. Rev. Lett.* **19**, 108.

Lieb, E.H. and Mattis, D.C. (1966). *Mathematical physics in one dimension.* Academic Press, New York.

Lieb, E.H. and Wu, F.Y. (1972). Two-dimensional ferroelectric models. In *Phase transitions and critical phenomena* (ed. C. Domb and M.S. Green) Vol. 1. Academic Press, New York.

Maxwell, J.C. (1874). On the dynamical evidence of the molecular constitution of bodies. *Nature, Lond.* **11**, 53. [Also *Collected works*, Vol. 2, p. 424. Dover, New York.]

Montroll, E.W., Potts, R.B., and Ward, J.C. (1963). Correlations and spontaneous magnetization of the two-dimensional Ising model. *J. math. Phys.* **4**, 308.

Onsager, L. (1939). Discussion remark at a conference on dielectrics. NY Academy of Sciences.

Onsager, L. (1944). Crystal statistics I: A two-dimensional model with an order–disorder transition. *Phys. Rev.* **65**, 117.

Onsager, L. (1949). Discussion remark: Spontaneous magnetization of the two-dimensional Ising model. *Nuovo Cim.* (*Suppl.*) **6**, 261.

Onsager, L. (1971). The Ising model in two dimensions. In *Critical phenomena in alloys, magnets and superconductors* (ed. R.E. Mills, E. Ascher, R.I. Jaffee). McGraw-Hill, New York.

Ornstein, L.S. and Zernike, F. (1914). Accidental deviations of density and opalescence at the critical point of a single substance. *Proc. Akad. Sci., Amsterdam* **17**, 793.

Pauling, L. (1935). The structure and entropy of ice and of other crystals with some randomness of atomic arrangement. *J. Am. chem. Soc.* **57**, 2680.

Pearce, P.A. and Thompson, C.J. (1978). The high density limit for lattice spin models. *Commun. math. Phys.* **58**, 131.

Peierls, R. (1936*a*). Ising's model of ferromagnetism. *Proc. Cambridge phil. Soc.* **32**, 477.

Peierls, R. (1936*b*). Statistical theory of superlattices with unequal concentrations of the components. *Proc. R. Soc., London* **A154**, 207.

Pippard, A.B. (1957). *Elements of classical thermodynamics.* Cambridge University Press, New York.

Potts, R.B. (1952). Some generalized order–disorder transformations. *Proc. Camb. phil. Soc.* **48**, 106.

Roller, D. (1950). *The early development of the concept of temperature and heat.* Harvard University Press, Cambridge, Massachusetts.

Ruelle, D. (1963). Classical statistical mechanics of a system of particles. *Helv. Phys. Acta.* **36**, 183.

Ruelle, D. (1969). *Statistical mechanics: rigorous results.* W.A. Benjamin, New York.

Rumford (Thompson, B.) (1798). An inquiry concerning the source of the heat which is excited by friction. *Phil. Trans. R. Soc., London* **80**.

Rushbrooke, G.S. (1963). On the thermodynamics of the critical region for the Ising problem. *J. Chem. Phys.* **39**, 842.

Rushbrooke, G.S. (1965). On the Griffiths inequality at a critical point. *J. Chem. Phys.* **43**, 3439.

Rys, F. (1963). Über ein zweidimensionales klassisches Konfigurations-modell. *Helv. Phys. Acta.* **36**, 537.

Schultz, T.D., Mattis, D.C., and Lieb, E.H. (1964). Two-dimensional Ising model as a soluble problem of many fermions. *Rev. mod. Phys.* **36**, 856.

Siegert, A.J.F. and Vezzetti, D.J. (1968). On the Ising model with long-range interaction. *J. math. Phys.* **9**, 2173.

Sinai, Ya.G. (1976). *Introduction to ergodic theory.* Princeton University Press.

Slater, J.C. (1941). Theory of the transition in KH_2PO_4. *J. Chem. Phys.* **9**, 16.

Smith, E.R. and Thompson, C.J. (1986). Glass-like behaviour of a spherical model with oscillatory long-range potential. *Physica* **135A**, 559.

Stanley, H.E. (1968). Spherical model as the limit of infinite spin dimensionality. *Phys. Rev.* **176**, 718.

Stanley, H.E. (1971). *Introduction to phase transitions and critical phenomena.* Oxford University Press, New York.

Sylvester, G.S. (1976). Inequalities for continuous-spin Ising ferromagnets. *J. stat. Phys.* **15**, 327.

Takahashi, H. (1942). Eine einfache Methode zur Behandlung der statistischen Mechanik ein-dimensionalen Substanzen. *Proc. Phys. Math. Soc., Japan* **24**, 60. [English translation: A simple method for treating the statistical mechanics of one-dimensional substances in Lieb and Mattis (1966), p. 25.]

Thompson, C.J. (1965). Algebraic derivation of the partition function of a two-dimensional Ising model. *J. math. Phys.* **6**, 1392.

Thompson, C.J. and Silver, H. (1973). The classical limit of *n*-vector spin models. *Commun. math. Phys.* **33**, 53.

Tonks, L. (1936). The complete equation of state of one, two, and three dimensional gases of hard elastic spheres. *Phys. Rev.* **50**, 955.

Uhlenbeck, G.E. and Ford, G.W. (1963). *Lectures in statistical mechanics.* American Mathematical Society, Providence, R.I.

Uhlenbeck, G.E., Kac, M., and Hemmer, P.C. (1963). On the van der Waals theory of the vapour–liquid equilibrium II: Discussion of the distribution functions. *J. math. Phys.* **4**, 229.

van Hove, L. (1949). Quelques propriétés générales de 1-intégrale de configuration d'un système de particules avec interaction. *Physica* **15**, 951.

van Kampen, N.G. (1964). Condensation of a classical gas with long-range attraction. *Phys. Rev.* **135A**, 362.

Vdovichenko, N.V. (1965). A calculation of the partition function for a plane dipole lattice. *Soviet Physics JETP* **20**, 477.

Widom, B. (1965). Equation of state in the neighborhood of a critical point. *J. Chem. Phys.* **43**, 3898.

Wilson, K.G. (1971). Renormalization group and critical phenomena I and II. *Phys. Rev.* **B4**, 3174, 3184.

Wilson, K.G. (1975). The renormalization group: Critical phenomena and the Kondo problem. *Rev. mod. Phys.* **47**, 773.

Wilson, K.G. (1983). The renormalization group and critical phenomena. *Rev. mod. Phys.* **55**, 583.

Wilson, K.G. and Fisher, M.E. (1972). Critical exponents in 3.99 dimensions. *Phys. Rev. Lett.* **28**, 240.

Yang, C.N. (1952). The spontaneous magnetization of a two-dimensional Ising model. *Phys. Rev.* **85**, 809.

Yang, C.N. and Lee, T.D. (1952). Statistical theory of equations of state and phase transitions I: Theory of condensation. *Phys. Rev.* **87**, 404.

INDEX